Hybrid Rough Sets and Applications in Uncertain Decision-Making

Systems Evaluation, Prediction, and Decision-Making Series

Efficiency of Scientific and Technological Activities and Empirical Tests
Hecheng Wu, Nanjing University of Aeronautics and Astronautics
ISBN: 978-1-4200-8846-5

Grey Game Theory and Its Applications in Economic Decision-Making
Zhigeng Fang, Nanjing University of Aeronautics and Astronautics
ISBN: 978-1-4200-8739-0

Hybrid Rough Sets and Applications in Uncertain Decision-Making
Lirong Jian, Sifeng Liu, and Yi Lin, Nanjing University of Aeronautics and Astronautics
ISBN: 978-1-4200-8748-2

Irregularities and Prediction of Major Disasters
Yi Lin, Nanjing University of Aeronautics and Astronautics
ISBN: 978-1-4200-8745-1

Optimization of Regional Industrial Structures and Applications
Yaoguo Dang, Nanjing University of Aeronautics and Astronautics
ISBN: 978-1-4200-8747-5

Systemic Yoyos: Some Impacts of the Second Dimension
Yi Lin, Nanjing University of Aeronautics and Astronautics
ISBN: 978-1-4200-8820-5

Theory and Approaches of Unascertained Group Decision-Making
Jianjun Zhu, Nanjing University of Aeronautics and Astronautics
ISBN: 978-1-4200-8750-5

Theory of Science and Technology Transfer and Applications
Sifeng Liu, Zhigeng Fang, Hongxing Shi, and Benhai Guo,
Nanjing University of Aeronautics and Astronautics
ISBN: 978-1-4200-8741-3

Hybrid Rough Sets and Applications in Uncertain Decision-Making

Lirong Jian · Sifeng Liu · Yi Lin

CRC Press
Taylor & Francis Group
Boca Raton London New York

CRC Press is an imprint of the
Taylor & Francis Group, an **informa** business
AN AUERBACH BOOK

Auerbach Publications
Taylor & Francis Group
6000 Broken Sound Parkway NW, Suite 300
Boca Raton, FL 33487-2742

Printed in the United States of America on acid-free paper
10 9 8 7 6 5 4 3 2 1

International Standard Book Number: 978-1-4200-8748-2 (Hardback)

Library of Congress Cataloging-in-Publication Data

Jian, Lirong.
Hybrid rough sets and applications in uncertain decision-making / Lirong Jian, Sifeng Liu, and Yi Lin.
p. cm.
Includes bibliographical references and index.
ISBN 978-1-4200-8748-2 (alk. paper)
1. Soft computing. 2. Data mining. 3. Rough sets. 4. Decision making. I. Liu, Sifeng. II. Lin, Yi, 1959- III. Title.

QA76.9.S63J53 2010
006.3--dc22 2010012957

Visit the Taylor & Francis Web site at
http://www.taylorandfrancis.com

and the Auerbach Web site at
http://www.auerbach-publications.com

Contents

Preface

Management activities consist of a series of decision-making processes. Faced with competitive markets in modern times, corporate managers and businessmen often encounter complex decision-making problems concerning uncertainty, and need to analyze and deal with various uncertainties such as random, fuzzy, preference, roughness, and grey. Besides, immediate decision is required. Decision-making under uncertain conditions is dealt with in different disciplines such as management science, information science, system science, computer science, knowledge engineering, reliability technology, and so on. Hybrids of soft computing technologies, such as rough set theory, grey system theory, fuzzy set theory, genetic algorithm, and neural network, can be used to flexibly deal with uncertainties in real life. The goal of soft computing technology is to exploit a fault-tolerant approach that can deal with imprecision, uncertainty, approximate reasoning, and partial correctness. Thus, a processable, powerful, and low-cost solution that is very similar to human decision-making can be achieved. The main objective is to design an effective and low-cost computing method by trying to derive an approximate solution.

Rough set theory, first introduced by a Polish scientist, Z. Pawlak, in 1982, has been described as a new mathematical tool to deal with vagueness and uncertainty. It can be effectively used to analyze information that is inaccurate, inconsistent, and incomplete. As a powerful approach to data reasoning, rough set theory has been successfully applied in knowledge acquisition; decision analysis and forecast; knowledge discovery from database, expert systems, and decision support systems; and in many other fields. Rough set theory and other soft technology theories, including grey system, artificial neural network, probability and statistics, and fuzzy set theory, are all complementary theories. The advantages of rough set theory can be blended with other soft technology theories to discover a more powerful hybrid method of soft decision-making. At the same time, the scope of rough set theory application can be broadened, and a wide range of scientific and standard methods for decision-making under uncertain conditions can be derived.

This book introduces the theory, method, and applications of rough sets. It systematically introduces the methods and applications of hybrid rough set theory, together with other related soft technologies, and also covers the latest research done by scholars from China and abroad. Its main aim is to address the uncertainties that could arise in the practical application of knowledge representation systems. The book consists of eight chapters: (1) "Introduction," (2) "Rough set theory," (3) "Hybrid of rough set theory and probability," (4) "Hybrid of rough set and dominance relation," (5) "Hybrid of rough set theory and fuzzy set theory," (6) "Hybrid of rough set and grey system," (7) "A hybrid approach of variable precision rough set, fuzzy set, and neural network," and (8) "Application analysis of hybrid rough set."

The main objectives of this book are to (1) emphasize the application of soft computing technology, (2) minimize the detailed explanation of complicated mathematical formulas, and (3) illustrate the specific application of hybrid methods to highlight rough set theory and other soft computing technologies.

This book can serve as teaching material for undergraduates and postgraduates majoring in economic management, applied mathematics, information science, automatic control, and it can also be used as a reference book not only for those majoring in humanities, social sciences, and other disciplines, but also for executives in enterprises and public institutions, engineering and technical personnel in scientific research institutions, and other researchers and practitioners. The book is a collaborative work of Professor Jian Lirong, Professor Liu Sifeng, and Professor Lin Yi. Doctor Liu Jiashu and Master Liu Yong have done a lot of translation for this book. We are sincerely thankful to them.

As rough set theory and other soft computing technologies are still in a process of development, and due to limitations in our knowledge, errors are unavoidable, and we welcome criticism and suggestions from experts and readers.

Acknowledgments

The research in this book has been supported by the National Natural Science Foundation of China (Nos. 90924022, 70971064, 70701017, and 70901041), the Key Project of Philosophic and Social Sciences of China (No. 08AJY024), the Key Project of Soft Science Foundation of China (2008GXS5D115), the Foundation for Doctoral Programs (200802870020), the Foundation for Humanities and Social Sciences of the Chinese National Ministry of Education (No. 09YJA630067) and the Foundation for Humanities and Social Sciences of the Chinese National Ministry of Education (No. 08JA630039), the Project of Soft Science Foundation of Jiangsu Province (No. BR2008099) the Key Project of Soft Science Foundation of Jiangsu Province (No. BR2008081), Soft Science Project of Defense Science and Technology Industry (No. 2008GFZC025) and the Foundation for Humanities and Social Sciences of Jiangsu Province (No. 07EYA017). At the same time, we would like to acknowledge the partial support of the Science Foundation for the Excellent and Creative Group in Science and Technology of Nanjing University of Aeronautics and Astronautics and Jiangsu Province (No. Y0553-091). While writing this book, we consulted widely and referred to research by several scholars. We wish to thank them all.

Authors

Lirong Jian received her PhD in management science and engineering from Southeast University, Nanjing, China, in 2004. She then had two years of postdoctoral experience specializing in management science and engineering at Nanjing University of Aeronautics and Astronautics, China. At present, she is serving as a professor at the College of Economics and Management of Nanjing University of Aeronautics and Astronautics; she is also working as a guide for doctoral students in management science and systems engineering.

Dr. Jian is principally engaged in forecasting and decision-making methods, soft computing, and project management and system modeling. She has also directed and/or participated in nearly 20 projects at the national, provincial, and ministerial levels, for which she received four provincial awards in scientific research and applications. Over the years, she has published over 40 research papers and 6 books.

Sifeng Liu received his bachelor's degree in mathematics from Henan University, Kaifeng, China in 1981, and his MS in economics and his PhD in systems engineering from Huazhong University of Science and Technology, Wuhan, China, in 1986 and 1998, respectively. He has been to Slippery Rock University, Pennsylvania, and to Sydney University, Australia, as a visiting professor. At present, Professor Liu is the director of the Institute for Grey Systems Studies and the dean of the College of Economics and Management of Nanjing University of Aeronautics and Astronautics. He is also a distinguished professor and guide for doctoral students in management science and systems engineering.

Dr. Liu's main research activities are in grey systems theory and in regional technical innovation management. He has directed more than 50 projects at the national, provincial, and ministerial levels, has participated in international collaboration projects, and has published over 200 research papers and 16 books. Over the years, he has received 18 provincial and national awards for his outstanding achievements in scientific research and applications. In 2002, one of his papers was recognized by the World Organization of Systems and Cybernetics as one of the best papers of its 12th International Congress.

Dr. Liu is a member of the evaluation committee of the Natural Science Foundation of China (NSFC) and a member of the standing committee for teaching guide in management science and engineering of the Ministry of Education, China. He also serves as an expert on soft science at the Ministry of Science and Technology, China. Professor Liu currently serves as the chair of the technical committee of the IEEE SMC on Grey Systems; the president of the Grey Systems Society of China (GSSC); a vice president of the Chinese Society for Optimization, Overall Planning and Economic Mathematics (CSOOPEM); a cochair of the Beijing Chapter and the Nanjing Chapter of IEEE SMC; a vice president of the Econometrics and Management Science Society of Jiangsu Province (EMSSJS); a vice president of the Systems Engineering Society of Jiangsu Province (SESJS); and a member of the Nanjing Decision Consultancy Committee. He serves as the editor in chief of *Grey Systems: Theory and Application*, and as a member of the editorial boards of over 10 professional journals, including *The Journal of Grey System* (United Kingdom); *Scientific Inquiry* (United States); *The Journal of Grey System* (Taiwan, China); *Chinese Journal of Management Science*; *Systems Theory and Applications*; *Systems Science and Comprehensive Studies in Agriculture*; and the *Journal of Nanjing University of Aeronautics and Astronautics*.

Dr. Liu has won several accolades, such as the National Excellent Teacher in 1995, Excellent Expert of Henan Province in 1998, National Expert with Prominent Contribution in 1998, Expert Enjoying Government's Special Allowance in 2000, Excellent Science and Technology Staff in Jiangsu Province in 2002, National Advanced Individual for Returnee and Achievement Award for Returnee in 2003, and Outstanding Managerial Personnel of China in 2005.

Yi Lin holds all his educational degrees (BS, MS, and PhD) in pure mathematics from Northwestern University, Xi'an, China and Auburn University, Alabama, and has had one year of postdoctoral experience in statistics at Carnegie Mellon University, Pittsburgh, Pennsylvania. Currently, he serves as a guest or specially appointed professor in economics, finance, systems science, and mathematics at several major universities in China, including Huazhong University of Science and Technology, Changsha National University of Defence Technology, and Nanjing University of Aeronautics and Astronautics, and as a professor of mathematics at the Pennsylvania State System of Higher Education (Slippery Rock campus). Since 1993, he has been serving as the president of the International Institute for General Systems Studies, Inc. Among his other professional endeavors, Professor Lin has had the honor of mobilizing scholars from over 80 countries representing more than 50 different scientific disciplines. Over the years, he has served on the editorial boards of 11 professional journals, including *Kybernetes: The International Journal of Cybernetics, Systems and Management Science*; the *Journal of Systems Science and Complexity*; the *International Journal of General Systems*, and *Advances in Systems Science and Applications*. He is also a coeditor of the book series entitled *Systems Evaluation, Prediction and Decision-Making*, published by Taylor & Francis (2008).

Some of Lin's research was funded by the United Nations, the State of Pennsylvania, the National Science Foundation of China, and the German National Research Center for Information Architecture and Software Technology. By the end of 2009, he had published nearly 300 research papers and over 30 monographs, and edited volumes on special topics. His works have been published by such prestigious publishers as Springer, Wiley, World Scientific, Kluwer Academic (now part of Springer), Academic Press (now part of Springer), and others. Throughout his career, Lin's scientific achievements have been recognized by various professional organizations and academic publishers. In 2001, he was inducted into the honorary fellowship of the World Organization of Systems and Cybernetics. Lin's professional career started in 1984 when his first paper was published. His research interests are mainly in the area of systems research and applications in a wide range of disciplines of traditional science, such as mathematical modeling, foundations of mathematics, data analysis, theory and methods of predictions of disastrous natural events, economics and finance, management science, and philosophy of science.

Chapter 1

Introduction

Data-mining and knowledge discovery have now become a very active research field, while rough set theory (RST) has been successfully applied to knowledge acquisition as a powerful tool for data reasoning, decision analysis and forecast, knowledge discovery from database, and expert system and decision support system. Because of high complementary between RST and other soft technologies, including grey system, artificial neural networks, probability and statistics, and fuzzy set theory, we can integrate the advantages of RST with other theories related to soft technologies and build a more powerful hybrid method of soft decision-making. Therefore, we can broaden the scope of RST application and provide a wide range of scientific and standard methods for decision-making under uncertain conditions.

1.1 Background and Significance of Soft Computing Technology

With the rapid development and wide application of the internet and database technology, data volume has been increasing at a striking speed, and a large amount of data is being applied in various fields of social life and industries. As a result, the traditional statistical techniques and data management tools are no longer adaptable to analyze these massive datasets. The massive data has been described as "rich data but poor knowledge." People need to adopt a higher degree automatic and more efficient data-processing approach to deal with large amounts of data and provide useful knowledge. From the financial sector to the manufacturing sector, more and more companies are dependent on the analysis of massive data to gain

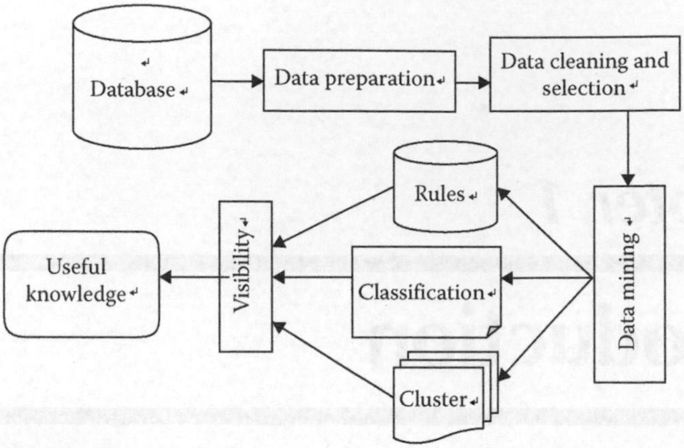

Figure 1.1 KDD process.

competitive advantage. Knowledge has become the primary driving force in social life and industry. To help people analyze huge amounts of data intelligently and automatically, a new generation of technologies and tools have emerged, which are mainly used in the field of data mining and knowledge discovery. KDD means automatically acquiring knowledge from the large databases, and aims at finding hidden, previously unknown, and potentially useful knowledge from the data. In essence, KDD tries to find the general rules and patterns from a large collection of data. Data mining can be viewed as a method to find these rules and models. The KDD process is shown in Figure 1.1.

1.1.1 Analytical Method of Data Mining

The analytical method of data mining can be functionally divided into five kinds: automatic prediction of trends and behavior, association analysis, cluster analysis, concept description, and deviation detection.

1.1.1.1 Automatic Prediction of Trends and Behavior

Data mining automatically searches for predictable information from large volume database. Although this process was previously done by a number of manual analyses, we can now quickly and directly draw conclusions just from the database. Market prediction is a typical example, in which data mining is done based on past promotion data to find a potential customer with the largest investment–return value in the future. Other cases of trend prediction include prediction of companies that can go bankrupt in the future and the selection of corresponding groups that are most likely to respond to the specified events.

1.1.1.2 Association Analysis

Data association is a class of important findable knowledge in the database. If there is certain regularity between two or more variable values, it can be called association. Association can be divided into simple incidence, time-series association, and causal association. The purpose of association analysis is to identify the hidden related networking in database. Sometimes people do not know the association function among data, and it is uncertain even if they know of that, so the generated rules from the association analysis with credibility also known as support is used to indicate the probability of their occurring. Shopping mall has a total of 1000 T business, of which 100 T bought a microwave oven and microwave oven–specific containers at the same time, the support of the association rules among buying microwave oven and microwave oven–specific containers at the same time is 10 percent.

The shopping basket analysis is a typical example of association rule mining. Market analysts need to find the relations among different commodities put into the customer's shopping basket from a large amount of data. If a customer buys milk, what is the possibility that he or she will buy bread? And which kind of group or collection of goods will customers be likely to buy at the same time? For example, 80 percent of customers buy milk and bread at the same time, or 70 percent of customers buy hammer and nails at the same time. These are the association rules extracted from the shopping basket data. The results of the analysis can help managers design different store layouts. One strategy is to place commodities that are frequently purchased at the same time close to each other, and in this way sales of these commodities can be further stimulated. For example, if customers buy a computer, they are likely to buy financial software at the same time, and it may help increase sales by displaying the hardware near the software. Another strategy is to place the hardware and software at two ends of the store, which may attract customers to purchase these products and select other commodities all the way.

1.1.1.3 Cluster Analysis

Records in the database can be divided into a series of meaningful subsets, and is known as clustering. Clustering enhances people's understanding of the objective reality, which is the prerequisite of concept description and deviation analysis. Clustering technologies mainly include traditional methods of pattern recognition and mathematical taxonomy. In the early 1980s, Mchalski put forward the concept of clustering technology, and the key point is that when the objects are divided, people not only consider the distance between objects, but also demand a connotative description between the divided content, so as to avoid a certain one-sidedness of traditional technologies.

1.1.1.4 Concept Description

It describes the connotation of certain types of objects and summarizes the relevant characteristics of the objects. It can be divided into characteristic description and differential description. The former describes the common characteristics of certain types of objects, while the latter describes the difference between the different types of objects. Generating a type of characteristic description only refers to the commonness among the objects.

1.1.1.5 Deviation Detection

There are often some abnormal records in the database, so it is significant to detect the deviations. Deviation includes much potential knowledge, such as the unusual example in the classification, a special case that does not meet the rule, the deviation between observations and model prediction, quantity value that changes over time, and so on. The basic method of error detection is to find the difference between observations and reference values.

1.1.2 Knowledge Discovered by Data Mining

The most common knowledge discovered by data mining is of the following five types:

1. *Generalization*. It refers to the general description of the characteristics of knowledge. According to the micro-features of data, finding representative and universal knowledge by means of high-level concepts and reflecting the common nature of similar kinds of things is, in summary, the refining and abstracting of data.
2. *Association*. It reflects the dependent or associated knowledge between an event and other events. If there is correlation among two or more attributes, then one of the attribute values can be predicted on the basis of other attributes.
3. *Classification*. It reflects the characteristic knowledge of the common properties shared by objects of the same type as well as the differentiated characteristic knowledge of different kinds of things.
4. *Prediction*. This knowledge is to predict future data based on the time-series data and according to the historical and current data, and viewed as the associated knowledge with time as the key attribute.
5. *Deviation*. It is the description of differences and extreme cases to reveal abnormal occurrence that deviates from the conventional, such as the special case of the standard categories and outlier value outside the cluster data. All the knowledge can be found from different levels of concept, so as to satisfy the needs of different users' levels of decision making with the upgrade of concept from micro to macro.

KDD is a typical cross-discipline that needs computer science, statistics, cognitive science, and other related knowledge to deal with real-world problems. In August 1989, KDD first appeared in the United States Eleventh International Joint Conference on Artificial Intelligence held in Detroit. In the subsequent 1991, 1993, and 1994 KDD symposia, researchers and application developers from various fields gathered to concenter and discuss data statistics, mass data analysis algorithms, knowledge representation, and knowledge application. The KDD Organizing Committee renamed the symposium as the International Conference, and held its first International Conference on KDD in Montreal, Canada. The Fourth Knowledge Discovery and Data Mining International Conference was held in New York in 1998, which not only carried out an academic discussion but also presented data mining software products produced by over 30 software companies, a lot of which have been applied in countries in North America and Europe. KDD and data-mining methods have currently become an active research area in the artificial intelligence and database technology. The uncertainty of multi-attribute decision-making has become an important component of modern decision-making science relying on the database. It is widely used in social, economic, management, engineering design, and many other fields. Some findings have been directly converted into a business plan, and greatly improve the quality of the company's decision-making. From the financial sector to the manufacturing sector, more and more companies are becoming dependent on the analysis of a massive amount of data to obtain a competitive advantage; hence, knowledge has become the primary driving force of social life. For example, the application of data mining helps to solve the decision-making in the placement of goods in shopping malls, of which the famous "beer and diapers" case is a successful example. Although different data mining methods have different objectives, generally speaking, they can be divided into two categories: verification and discovery. Verification is limited to prove users' hypothesis, while discovery is used to search for new models. Discovery can be divided into prediction and description. Prediction models found through the system help to guide the future behavior, while description can provide user models in an understandable form. Data mining usually needs to deal with incomplete and uncertain large-capacity data, which has similarity with other disciplines of applied science. Facing the complex uncertainty of decision-making, it is impractical to try to accurately characterize with a complete mathematical model; even though it is feasible, it is also very difficult to solve and analyze. To help people intelligently analyze data and automatedly study a number of cases, there has been a new generation of soft computing tools, such as RST, the grey system theory, fuzzy set theory, genetic algorithm, and neural network. These soft computing technologies and the hybrid of their advantages are aimed to exploit inaccuracy, uncertainty, approximate reasoning, and partial correctness in the process of decision-making, so as to obtain processable, powerful, and low-cost solutions that are very similar to human decision-making.

Many problems in data mining and knowledge discovery can be coagulated into the decision-making of uncertainty in management activities. Management activities are comprised of a series of decision-making, especially in today's fiercely competitive market; whether businesses or individuals, they often encounter complex uncertainties such as randomness, fuzziness, preferences, roughness, greyness, etc., that need to be analyzed and to be dealt with in decision-making, and make rapid decisions. Uncertainty in decision-making generally exists in many areas such as management science, information science, systems science, computer science, knowledge, technology, engineering, and reliability technology. Because many uncertainties in the real world are so complicated that no analytic and accurate models can be used to describe them and precise measurement and controlled hard technology are not always effective in dealing with such complex issues, there is a need to rely on human intuition, which contains intuition based on human thinking and achieved subjective thinking, such as a case in which experts recognize a human face against the background of noise. RS, grey systems, probability theory, fuzzy set and neural network, and other soft computing technologies match the reasoning in human thinking and humans' extraordinary capacity of learning, taking full advantage of the human intuitive knowledge, which are effective means to deal with uncertainties in decision-making. Compared with the accurate, certain, and strict hard computing technologies, soft computing technologies are effective in obtaining inaccurate or suboptimal but economical solutions, and they can compete hard with computing technologies. The unique features of soft computing technologies have attracted the interest of a variety of academic research teams.

1.2 Characteristics of Rough Set Theory and Current Status of Rough Set Theory Research

In many KDD problems, learning category is a kind of problem that has been researched extensively. As a powerful tool for data reasoning, RST is a theory to deal with imprecise and uncertain and incomplete data, and effectively analyze various deficient information that is inaccurate, inconsistent, and incomplete. RST was originally proposed by the Polish scientist Z. Pawlak in 1982. Because the original research of the RST was published mostly in Polish, it did not attract scholars of international computer science and mathematics to give more attention to this theory and was only studied in Eastern European countries, but it was not until the end of the 1980s that this theory aroused the increasing attention of scholars all over the world. In 1992, the first international academic conference on RST was held in Poland. In 1995, ACM Communication listed it as a new emerging research topic in computer science. In 1998, the *International Journal of Information Science* (*Information Sciences*) published an album for the RST. The RST has been studied and applied around in mathematics, computer, and various specialized fields.

1.2.1 Characteristics of the Rough Set Theory

The success of the RST is mainly due to the following advantages: it only depends on the raw data, without any external information; RS method not only applies to analyze the attributes of quality but also to study the attributes of quantity; it has the attributes of reduce redundance, and algorithm of attribute reduction is relatively simple and this can be expressed by the minimal knowledge given by the decision-making rules that are derived from the RS model; it does not require amendments to the inconsistencies, and it can divide the inconsistent rules into the certain rules and uncertain rules; and the results derived from RS methods are easy to understand.

The data collected from the real world may contain all kinds of noise, and there are many uncertainties and incomplete information to be dealt with. The traditional methods of dealing with the uncertain information such as fuzzy set theory, evidence theory, the theory of probability, and so on, need additional information or a priori knowledge of the data, so it is hard for them to deal with a large amount of data in the database aspect. As a soft computing method, the significant difference between RST and other theories dealing with uncertainty and imprecision is that it is not required to provide any a priori information except the data collection that are needed to deal with the problem, such as probability distribution in the statistics and the membership degree in the fuzzy set theory, so it can be said to be objective in describing or dealing with the uncertainty of the problem.

RST is a new mathematical method that is driven by practical needs for data analysis and data mining. RST and KDD are closely related; the former provides a new method and tool for KDD, due to the following reasons:

1. The object that KDD studies is mostly the relational database, and the relationship table can be seen as a form of decision table in RST, while this gives the application of RS method a great convenience.
2. There are certain rules in the real world. Discovering the uncertain information from the database provides a place for RS to apply.
3. Discovering the abnormal situation from the data and ruling out the disturbance of noise in the process of knowledge discovery is also a strength of the RS method.
4. Using the knowledge discovery algorithm with the RS method can be in favor of the parallel implementation and improve mining efficiency greatly. For the large-scale knowledge discovery in the databases, it is very important.
5. Using RS method to do pretreatment and remove redundant attributes can improve mining efficiency and reduce the error rate.
6. Compared with the fuzzy set method or neural network method, the decision-making rules and the reasoning process from the RS method are much easier to be confirmed and explained.

1.2.2 Current Status of Rough Set Theory Research

RST can be divided into two categories in the study of artificial intelligence: decision-making analysis and non-decision-making analysis.

1.2.2.1 Analysis with Decision-Making

Analysis with decision making is mainly applied to supervised learning and decision-making analysis. The applications of RST in supervised learning can be divided into two aspects: do preprocessing for the training set of learning and get the rules with the applications of RS method.

Taking into consideration redundant attributes usually contained in the obtained training sets due to the actual measurement, performing preprocessing is necessary for the training set of learning, and the application of routh set attribute reduction can effectively remove redundant attributes. For example, with RS doing the data for pea disease can reduce the number of original attributes from 35 to 9, while analyzing the voting data of the United States House of Representatives in 1984 reduced the number of original attributes from 16 to 9. There is also redundancy for each attribute, so the application of the reduction technology of the RS method can delete the excess value of some attributes; in addition, after doing preprocessing with the application of RS for neural network's data, the number of the condition attributes and the value of attributes will be simplified.

Acquiring rules from the application of RS is a typical supervised learning, which used the value reduction method derived from the RS to acquire the rules directly from crude set. Because all the value reduction acquired directly from the decision-making table has been proved to be an NP-complete problem, using the heuristic algorithm that is neighborhood independent to solve a minimum reduction is a commonly used method. In the application of decision analysis that uses reduction attribute, value reduction and nuclear and other concepts of the RST can provide the most useful information to reduce and research the decision-making information.

1.2.2.2 Non-Decision-Making Analysis

Analysis of non-decision-making mainly includes data compression, reduction, clustering, pattern discovery, machine discovery, and so on. Non-decision-making data analysis is mainly that removing unnecessary attributes by taking advantage of reduction attribute and compressing data by using value reduction can do cluster analysis for the data. Because data reduction of the non-decision-making is also an NP-complete problem, there is still a need to solve a minimum reduction with the heuristic knowledge. Machine learning belongs to one branch of the typical artificial intelligence applications; especially for discovering knowledge in a large-scale database, rough set is considered as a very effective method.

At present, a lot of knowledge discovery systems based on RS have been developed, which include LERS (learning from examples based on RSs) developed by Kansas University, the United States, in which there are two different methods used to acquire rules, one of which is to account for enough rules set by using a method of machine and the other is to calculate all the rules set through the knowledge acquisition; the ROSE system developed by the Intelligent Decision Support System Laboratory of Computer Science Institute of Poznan University of Technology, Poland, which provides not only all RST's basic computing but also a number of similar techniques avoiding discrete data, such as similar relationship and advantage relationship, and these technologies are simple to be controlled by all users; the KDD-R system developed by the University of Regina, Canada; and the Rough Enough system and Rosset system developed by Norwegian Troll Data Inc.

RST has aroused the attention of Chinese research scholars since the beginning of the 1990s with papers on RS appearing after 1997. In recent years, research on RS has become gradually more and more popular, and the *Chinese Journal Network* has published more than 3000 articles on the RS covering many disciplines and applied fields. Different disciplines and areas focus on different aspects; for example, the mathematics field focuses on the theory and extension of RS method and the computer industry focuses on the algorithm design of the RS method and other fields focus on the application of RST.

1.3 Hybrid of Rough Set Theory and Other Soft Technologies

Compared with other theories in dealing with uncertainty, RST has some irreplaceable advantages, but still has some one-sidedness and shortcomings. For example, it greatly reduces the new predictive capability due to the over-fitting for data and cannot deal with the classification of multi-attribute decision-making; the border region that the RS depicts is relatively simple, such as equivalence relations based on the classification of RS is certain, while there is not belonging or containing to a certain extent; it can identify the random rule only supported by a few examples; it has no appropriate approach to deal with the raw data which are ambiguity.

According to probability statistics made by some scholars, each method has its own scope of application, and there is no method that can solve all the problems. In practical application, we often combine several techniques to construct a "hybrid" approach, so that it could complement each other to overcome the limitations of individual techniques, avoiding the existing method's disadvantages or weaknesses when used separately; such a hybrid system is superior to the use of a single method, and the "hybrid" in the paper means combining with some advantages of the existing methods. Another function of hybrid is to demonstrate various information processing in a system structure, and the hybrid of a number of means is a promising direction of development of KDD and data mining.

RST and the artificial neural network, probability theory, fuzzy set theory, the grey system theory, and other soft computing technologies are highly complementary to each other, and some scholars have explored the RS and artificial neural networks, probability theory, the grey system, fuzzy set theory, and the theory, method, and application that combines two or more soft technologies.

1.3.1 Hybrid of Rough Sets and Probability Statistics

As a measurement of random events, probability statistics reflects a kind of uncertainty, and it plays an important role in the application of uncertain reasoning. There are two understandings toward probability. One is interpreted as the degree of confidence reflecting people's experience and knowledge and it is called "subjective probability," while the other is understood as the relative frequency that appears after a large amount of repeated experiments, and known as the "objective probability." Subjective probability and objective probability can both reflect the regularity that meet certain kinds of statistics, and are both objective. However, the classical statistical methods can only be applied to the large sample data analysis, and are dependent on the events that have happened and need prior probability distribution.

In the information systems, there are generally two kinds of knowledge in the repository: one with all the description of the objects being completely known, such as Pawlak RS model; the other with only part of the description known being uncertain, so we can only describe the concepts through the information provided by training samples. To make the rule acquired from the training samples suitable for all objects in the discussing region, we should obey the rules in statistics, while we could ignore it if RST is applied, so it appears natural for statistics as a subject that does research in the rules of enormous amount of random phenomena in nature, human society, as well as the technical processes to be hybridized with RST. The hybrid of RSs and probability statistics can expand the functions of RS method, and also acquire the probability decision-making rules from the noisy data. Ziarko hybridized the RSs and probability statistics and put forward the variable precision RS model. Being an extension of the original RST, the variable precision RS loosens the strict definition of RST toward the approximate boundaries through setting the threshold value β, and is an allowable probabilistic classification. Compared with the standard RS model, there is a confident degree in the correct classification of objects in the variable precision RS model, which, on one hand, completes the concept of approximate space and, on the other hand, is beneficial to find correlative information through seemingly irrelevant data according to the RST.

1.3.2 Hybrid of Rough Sets and Dominance Relation

The classical RS methods cannot detect the inconsistency related with the preference, such as the price, fuel amount, speed, etc., and these attributes involve preference information but they are not considered in RS. If we replace the indiscernibility relation with the dominance relation to restructure the RS models, we can extend RS models as methods to solve problems met to analyze preference with multiple attributes, and these hybrid models have not only the best quality of classical RS models but also the ability to deal with the possible inconsistency that exists in analyzing preference with multiple values as well as making decisions related with typical cases, so that we can create rules that are more easily understood by users with the usage of hybrid models like this. Besides, rules based on the dominance relation are much more suitable than those based on the indiscernibility relation when they are used to classify new objects. Because we may often meet preference information when handling with economic, managing, or financial decision-making problems, the hybrid of RSs and dominance relation can enlarge the usage of RS models in economic, management, or financial fields. Greco and other scholars have put forward RS model and its extended models based on dominance relation by replacing the indiscernibility relation with the dominance relation, and these models can do well with the possible inconsistency that exists in analyzing preference with multiple attributes as well as making decisions related with typical cases.

1.3.3 Hybrid of Rough Sets and Fuzzy Sets

Fuzzy set theory is a mathematical tool that simulates the fuzziness in human classifying systems, and it can be used to obtain and simulate and even reason the fuzziness in practical information by mainly studying the fuzzy information granularity. Due to its simplicity and the similarity to human thoughts, the concept of fuzzy set is often used to express the quantitative data that is in language, such as "it's warm today," "he is honest," and "the oil price may rise dramatically in the near future," and it can also be used as the membership function in an intelligent system. The fuzzy set cannot be described with any precise mathematical formula and included in the processes of human thinking and the physiology, because any precise mathematical formula cannot be applied in the processes of the human reasoning physiology, and fuzzy sets are very important in classifying modes. However, standard fuzzy sets are defined by $U \rightarrow [0,1]$; although fuzzy system based on the definition can provide a good standard mathematical model, there is lack of the fuzzy membership value and semantic explanations of the operation of fuzzy set theory. Besides, fuzzy set theory does not have a mechanism from data learning; instead, the knowledge should be given clearly by the designers in the applied fields, which would bring some difficulties for the application of fuzzy set theory.

The fuzzy set theory and RST extend the classical set theory in solving problems of uncertainty and imprecision. Although there is some compatibility and similarity between them, they focus on different aspects. First, as to the description of knowledge granularity, fuzzy sets are described approximately through the membership degree of sets with regard to the objects, while RSs are described through a pair of the upper and lower approximation of some usable knowledge representation system with regard to the sets. Second, as to the relation of objects in the sets, fuzzy sets emphasize the ill-definitions of set boundary, known as the disputability of boundary, while RSs emphasize indiscernibility among different objects in a set; Last, with regard to the research objects, fuzzy sets study the membership relations of different objects in a same type, while RSs study the set relations consisting of different types of objects, so the former focuses on the degree of membership and the latter on the classification. Although the fuzzy membership function and the rough membership function both reflect the fuzziness of a concept and thus there is, to some extent, similarity between them intuitively, the former is mostly given by experts based on their experience, while the latter is directly obtained from the analyzed data. That is to say, the latter is more subjective. When the two theories are applied in fuzzy, uncertain, or incomplete information systems, they are different and uncompetitive. On the contrary, they are complementary naturally, so some "hybrid" of the two theories can be used to describe the uncertainty and imprecision, and take on a more powerful function. The "hybrid" of the two theories can be regarded as a method to simulate basic fuzzy granularity, in which intuition has fuzzy boundary and granularity attribute value. Some scholars like Dubios and Prade made hybrids of fuzzy sets and RSs, and put forward rough fuzzy set and fuzzy RS, which, respectively, are used to solve the problem that the concept of knowledge module is clear in the knowledge base while the concept described is vague, and that the problem is the concept of knowledge module is vague in the knowledge base while the concepts described are clear. Slowinski, Pal, and other scholars moved the research focus to the concept of similarity, which made the two concepts closer to each other in the expanded version.

1.3.4 Hybrid of Rough Set and Grey System Theory

Grey system theory was founded by Professor Deng Julong, a well-known Chinese scholar, in 1982, and can be regarded as a production of views and methods in cybernetics extended to social and economic systems, or a peripheral subject combined with automatic control science and the mathematic methods in operational research, while its content includes modeling, prediction, decision-making, and control. The grey system theory introduced is described as a mathematical tool to handle small sample and poor information. The systems with partially known and partially unknown information are viewed as grey systems, and valuable information is extracted or produced by mining the partially known information, so that the behavior and the law of system can be properly described and effectively controlled. Its characteristic is "modeling with few data." One important part of grey system is "grey prediction and decision-making," which is widely applied to many fields. The grey prediction has played a key role in the prediction of many important programs and its predicting precision is matchless. However, some of the mathematic basis of the grey theory is not precise enough. For example, there are some shortcomings in definition of grey degree and grey number calculation, and it is hard to find out a reasonable explanation for the outcome of grey number calculation, grey degree of the sum, margin, product, and quotient between two grey numbers, or the relations between different grey degrees in every grey number. These factors result in a severe lag in the research on grey equations and grey matrixes, and the existing inexplicit grey numbers in the general model (GM) of the grey system.

RST and grey system theory have a certain degree of relevance and complementarity. Both of them make it possible to find a model from the data made fuzzy by too many details through reducing the precision of data expression. At the same time, neither of them require any transcendental knowledge. RST deals with nonoverlap of the sort of roughness and the rough concept and focuses on the indiscernibility between objects, while grey system theory deals with the grey fuzzy sets that are "explicit extension but ambiguous intension" and emphasizes the uncertainty of poor information. The rough membership degree of RSs is directly acquired by calculating from the data analyzed, and it can depict objectively the uncertainty of the data. The hybrid models of RS and grey system have been discussed by some scholars, which can make up for the shortcomings in the definitions of the two theories. For example, a rough membership function can be used to complement the shortcomings of the definitions of the grey degree of grey numbers, while knowledge of different grey grades can be obtained from grey degree of grey numbers. Besides, we can compare the relative dominating degree of reduction attribute by applying the associated grey degree.

1.3.5 Hybrid of Rough Sets and Neural Networks

Neural network is composed of a large amount of highly interrelated processing elements (nerve cells), which would solve certain problems by coordinated work. Neural network can distill valuable and precise modes from complicated and (or) imprecise data, and is a kind of artificial intelligent technology widely used in identifying modes and machine learning. With a relatively high adaptability and learning ability, its precision in the data classification and predicting is very high. The difference between neural networks and classical statistical methods lies in the fact that the former can be applied in the data analysis of a small amount of samples without any transcendental knowledge. What's more, neural networks can do well in simulating data with complex modes because of the nonlinear transformation comprised in the recessive layer. These features make neural network system a strong functional method to simulate human action. However, neural network for calculation is very complex, and in most cases it would take much more time for neural networks to learn the calculating methods than the other ways in artificial intelligent technology. In some cases, the training time of neural networks is even measured in hours or days. Besides, it is difficult for neural networks to provide a clear description how the modes are found, so it is usually applied to the tasks that needn't be explained.

With a relatively high adaptability, capability for errors, and universalizing ability, neural network methods can make up RSs' shortcomings like the sensitivity toward noise data and the weakness at universalizing ability. On the other hand, RS methods can help make up the weak points that exist in neural network methods, such as difficulties in ascertaining relatively important attribute combinations, lacking general ways to build a network structure, opaque reasoning process, and lack of explanation ability. The hybrid of RS and neural networks may help to extract rough neighborhood knowledge that can be used to describe different concepts or categories in a regular style, and code them as network parameters; therefore, we can build the original repository network that can be used to learn more effectively. Substantively, the hybrid of RS and neural networks is a combination of two thinking manners—logical thinking and visualizing thinking. This kind of hybrid makes it possible to discover knowledge and make predictions from a fuzzy, incomplete, and noisy database, and to include the structural knowledge in the system. This ability is vital for data analysis in business, because there are many incomplete data in commercial fields, and we need to filter useful information from a large amount of noise data input, but sometimes it's expensive to collect unnecessary data when the network is applied in practice, and these unnecessary data may even disturb the explanation of network toward the final predicting results. Some scholars have already taken RS as the pretreatment tool of data to study the methods and application of the hybrid of RS and neural network, which would accelerate the learning speed of networks and improve its predicting ability as well as its explaining ability toward the data.

1.4 Summary

RS, grey system, probability theory, fuzzy set and neural network, and other soft technologies are able to take advantage of the intuitive knowledge of mankind to deal with uncertainty and simulate humans' decision-making. For the applications of the soft computing technologies in many uncertain decision-making, on the whole, most of them are still art, and the mechanisms are different for each soft technology to simulate human thinking and learning capabilities. On the basis of analyzing of the characteristics of RST and the present situation and the existing problems, and aiming at using knowledge to represent system that may contain a variety of uncertainties, this chapter describes the soft computing technology's background and significance and presents generally the hybrid between RST and other soft technologies. The reason why different soft technologies need to be hybridized together is mainly that it is difficult to satisfy the needs with regard to some certain areas, data sets, or system life-cycle stages. For example, it is difficult to identify which method can give a satisfactory result in some application fields' decision-making categories. Therefore, comparing and merging a number of technologies is a reasonable solution. In addition, the hybrid of various means helps explain the result of data mining. Because RST and other soft computing technologies can be complementary, this hybrid system is superior to the use of a single technology and can solve many urgent uncertain decision-making problems, and the accessible valuable knowledge can not only be used to guide practice, but also provide recommendations to the decision maker to improve decision-making and management capabilities.

Chapter 2

Rough Set Theory

Rough set theory is a kind of mathematical method driven by practical needs to deal with vagueness and uncertainty. The main idea of rough set theory is based on the indiscernibility relation that every object is associated with a certain amount of information and the object can only be expressed by means of some obtained information. Therefore, objects with the same or similar information can be indiscernible with respect to the available information. The indiscernible blocks formed by indiscernible objects of the universe are called the element sets or the knowledge granularity. Based on the knowledge used in approximation, the universe is divided into some element sets of the indiscernible objects that are based on conditional attribute sets, and the element sets are regarded as approximate "knowledge granularity"; the knowledge used in approximation would divide the universe into the decision-making class derived from the decision attribute sets, on this basis, and one kind of knowledge can be used to approximate the other. When approximation is inaccurate, roughness would appear.

2.1 Information Systems and Classification

The starting point of rough set is a dataset, which is usually organized into a table, and it is called information systems or databases. The basic operators in rough set theory are lower and upper approximations, and they are used to define the total dependence and the part dependence of the property in the databases. Data reduct is a very important concept in rough set theory.

2.1.1 Information Systems and Indiscernibility Relation

Knowledge representation is realized by information systems in the rough set model, and the form of information systems is the two-dimensional table showing the relationship between objects and values of attribute, which is similar to the relational database. When classification is conducted according to the characteristics of the object, each attribute value is associated with one group of value set, and when a value of each attribute is selected, a description of an object is given.

Definition 2.1 Suppose that $S = (U, A, V, f)$ is an information system, and also can be called the knowledge representation system, where $U = \{U_1, U_2, \ldots, U_{|U|}\}$ is a finite nonempty set, and named universe object space; $A = \{a_1, a_2, \ldots, a_{|A|}\}$ is the finite nonempty set of attributes. $V = \cup V_a$, where $a \in A$, V_a is the domain of the attribute a; and $f: U \times A \to V$ is an information function, for $\forall a \in A$, $\forall x \in U$, and $f(x, a) \in V_a$, it appoints the attribute value of each object in U.

Definition 2.2 Suppose $\forall a \in A$, $\forall x \in U$, and $f(x, a) \in V_a$; to every subset $\phi \neq P \subseteq A$, its indiscernibility relation I in U is defined as

$$I = \{(x, y) \in U \times U : f(x, q) = f(y, q) \forall q \in P\}$$

If $(x, y) \in I$, then x and y are indiscernible. Obviously, such definition of the indiscernibility relation is an equivalence relation, that is to say, the indiscernibility relation is characterized by reflexivity, symmetry, and transitivity. The equivalence class containing the object x is marked as $I(x)$. The equivalence class of indiscernibility relation is called as the element set or the atom in S, while the empty set is also the element of approximation space S. Equivalence class corresponds to the expressions of knowledge granularities, which is the basis of the main concepts in rough set, such as approximation, dependence, reduct, and so on.

2.1.2 Set and Approximations of Set

The definition of set in the rough set theory is relevant with the available information (knowledge) and the understanding of the relevant universe elements. In other words, we can see universe elements with the available information of the universe elements. Therefore, the two elements described by the same information are indiscernible.

Rough set method is to make a more identifiable data model based on the assumption of reducing the accuracy in data representation. The core assumption of "rough set" is that the knowledge is embodied in the ability of classification. That is to say, the rough set method is regarded as a standardized framework to find facts from incomplete data, whose result is expressed by the classification models or rules obtained from example sets.

Rough set theory embeds the knowledge to be classified into a set and makes it a part of the set. According to existing knowledge, there are three different situations as to whether an object $x \in U$ belongs to the set $X \subseteq U$, and they are as follows: (1) an object x absolutely belongs to X; (2) an object x absolutely does not belong to X; (3) an object x may belong to X and may not belong to X. On the basis of the above analysis, we put forward the two most important concepts in rough set theory: lower approximation and upper approximation.

Definition 2.3 Given a knowledge representation system $S = (U, A, V, f)$, $P \subseteq A$, $X \subseteq U$, $x \in U$, the lower approximation, upper approximation, negative region, and boundary of set X regarding I, respectively, are

$$\underline{apr}_P(X) = \bigcup\{x \in U : I(x) \subseteq X\} \tag{2.1}$$

$$\overline{apr}_P(X) = \bigcup\{x \in U : I(x) \cap X \neq \phi\} \tag{2.2}$$

$$neg_P(X) = \bigcup\{x \in U : I(x) \cap X = \phi\} \tag{2.3}$$

$$bnd_P(X) = \overline{apr}_P(X) - \underline{apr}_P(X) \tag{2.4}$$

The lower approximation of set X is actually the largest union of all the objects that surely belong to X, and is also called the positive region of X denoted as $pos_P(X)$; the negative region of X $neg_P(X)$ is the set that is combined by all the objects that definitely do not belong to X, thus $neg_P(X) = U - \overline{apr}_P(X)$; the upper approximation $\overline{apr}_P(X)$ of X is consists of all nonempty sets of equivalence classes that intersect with X, that is to say, it is the smallest set that consists of the objects that may belong to X, apparently, $U = \overline{apr}(X) \cup neg(X)$, while the boundary of X $bnd_P(X)$ is the difference between the upper approximation and lower approximation of the set X. The boundary region cannot be judged, for example, in terms of the attribute set P; there is no object in boundary region that can be classified into X or $-X$. If the $bnd_P(X)$ is an empty set, then the set X is a exact set with respect to I; otherwise, if the $bnd_P(X)$ is not an empty set, then the set X is rough set with respect to I. The sketch map for the concept of rough set is shown in Figure 2.1, from which we can see the concept of lower approximation, upper approximation, boundary region, and so on. The sketch describes the approximate properties of the set that cannot be defined exactly.

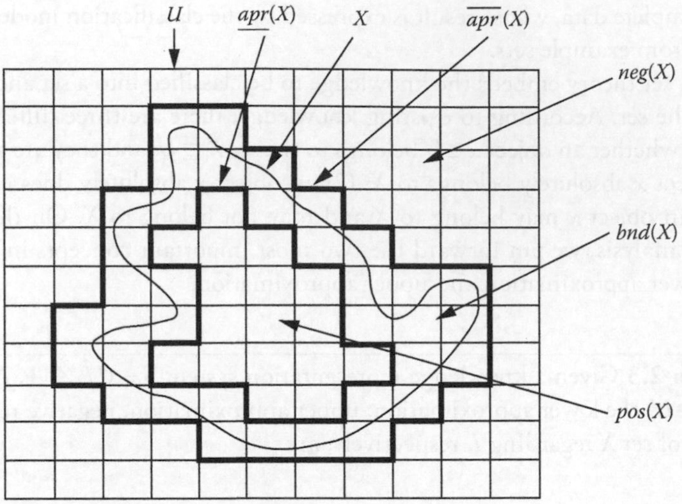

Figure 2.1 Sketch map for concepts of rough theory set.

Proposition 2.1 The properties of lower approximation and upper approximation are as follows:

1. $\underline{apr}_P(X) \subseteq X \subseteq \overline{apr}_P(X)$
2. $\underline{apr}_P(U) = \overline{apr}_P(U) = U$
3. $\underline{apr}_P(\phi) = \overline{apr}_P(\phi) = \phi$
4. $\overline{apr}_P(X \cup Y) = \overline{apr}_P(X) \cup \overline{apr}_P(Y)$
5. $\underline{apr}_P(X \cup Y) \supseteq \underline{apr}_P(X) \cup \underline{apr}_P(Y)$
6. $\overline{apr}_P(X \cap Y) = \overline{apr}_P(X) \cap \overline{apr}_P(Y)$
7. $\underline{apr}_P(X \cap Y) \subseteq \underline{apr}_P(X) \cap \underline{apr}_P(Y)$
8. $\underline{apr}_P(-X) = -\overline{apr}_P(X)$
9. $\overline{apr}_P(-X) = -\underline{apr}_P(X)$
10. $\overline{apr}_P(\overline{apr}_P(X)) = \underline{apr}_P(\overline{apr}_P(X)) = \overline{apr}_P(X)$
11. $\underline{apr}_P(\underline{apr}_P(X)) = \overline{apr}_P(\underline{apr}_P(X)) = \underline{apr}_P(X)$

Proof

1. Suppose $x \in \underline{apr}_P(X)$, then $I(x) \subseteq X$, while $x \in I(x)$, thus $x \in X$, $\underline{apr}_P(x) \subseteq X$. For $x \in X$, then $I(x) \cap X \neq \phi$, thus $x \in \overline{apr}_P(X)$, $X \subseteq \overline{apr}_P(X)$.
2. For $\underline{apr}_P(U) \subseteq U$ and $\forall x \in U$, then $I(x) \subseteq U$, thus $x \in \underline{apr}_P(U)$, $U \subseteq \underline{apr}_P(U)$, $\underline{apr}_P(U) = U$. For $\overline{apr}_P(U) \supseteq U$, thus $\overline{apr}_P(U) \subseteq U$, $\overline{apr}_P(U) = U$.
3. For $\underline{apr}_P(\phi) \subseteq \phi$, while $\phi \subseteq \underline{apr}_P(\phi)$, thus $\underline{apr}_P(\phi) = \phi$.

   Suppose $\overline{apr}_P(\phi) \neq \phi$, then $x \in \overline{apr}_P(\phi)$, that is to say, $I(x) \cap \phi \neq \phi$ while $I(x) \cap \phi \neq \phi$ which is contradictive with the assumption, thus $\overline{apr}(\phi) = \phi$.
4. Suppose $x \in \overline{apr}_P(X \cup Y)$, then $I(x) \cap (X \cup Y) \neq \phi$, thus we can get $(I(x) \cap X) \cup (I(x) \cap Y) \neq \phi$, $I(x) \cap X) \neq \phi$ or $(I(x) \cap Y) \neq \phi$, thus $x \in \overline{apr}_P(X)$ or $x \in \overline{apr}_P(Y)$, namely, it is $\overline{apr}_P(X \cup Y) = \overline{apr}_P(X) \cup \overline{apr}_P(Y)$.
5. For $X \subseteq X \cup Y$ and $Y \subseteq X \cup Y$, then $\underline{apr}_P(X) \subseteq \underline{apr}_P(X \cup Y)$ and $\underline{apr}_P(Y) \subseteq \underline{apr}_P(X \cup Y)$, thus $\underline{apr}_P(X) \cup \underline{apr}_P(Y) \subseteq \underline{apr}_P(X \cup Y)$.
6. For $X \cap Y \subseteq X$ and $X \cap Y \subseteq Y$, then $\overline{apr}_P(X \cap Y) \subseteq \overline{apr}_P(X)$ and $\overline{apr}_P(X \cap Y) \subseteq \overline{apr}_P(Y)$, thus $\overline{apr}_P(X \cap Y) \subseteq \overline{apr}_P(X) \cap \overline{apr}_P(Y)$.
7. Suppose $x \in \underline{apr}_P(X \cap Y)$, then $I(x) \subseteq X \cap Y$. From this we can get $I(x) \subseteq X$ and $I(x) \subseteq Y$, that is $x \in \underline{apr}_P(X) \cap \underline{apr}_P(Y)$, thus $\underline{apr}_P(X \cap Y) = \underline{apr}_P(X) \cap \underline{apr}_P(Y)$.
8. Suppose $x \in \underline{apr}_P(X)$, then $I(x) \subseteq X$, from which we can get $I(x) \cap -X = \phi$, that is the same as $x \notin \overline{apr}_P(-X)$, $x \in -\overline{apr}(-X)$, thus $\underline{apr}_P(X) = -\overline{apr}_P(-X)$.
9. Suppose $x \in -\underline{apr}_P(X)$, then $I(x) \subseteq -X$, from which we can get $I(x) \cap X = \phi$, namely $x \notin \overline{apr}_P(X)$, $x \in \overline{apr}(-X)$, thus $\overline{apr}_P(-X) = -\overline{apr}_P(X)$.
10. Due to $X \subseteq \overline{apr}_P(X)$, then $\overline{apr}_P(X) \subseteq \overline{apr}_P(\overline{apr}_P(X))$, and suppose $x \in \overline{apr}(\overline{apr}(X))$, then $I(x) \cap \overline{apr}_P(X) \neq \phi$, thus there exist $y \in I(x)$ and $y \in \overline{apr}_P(X)$, that is to say, $I(y) \cap X \neq \phi$. For $I(y) = I(x)$, then $I(x) \cap X \neq \phi$, $x \in \overline{apr}_P(X)$, then we can get that $\overline{apr}_P(X) \supseteq \overline{apr}_P(\overline{apr}_P(X))$, thus $\overline{apr}_P(\overline{apr}_P(X)) = \overline{apr}_P(X)$.

    From formula (2.1), we can get $\underline{apr}_P(\overline{apr}_P(X)) \subseteq \overline{apr}_P(X)$; suppose $x \in \overline{apr}_P(X)$, then $I(x) \cap X \neq \phi$, so we can get $I(x) \subseteq \overline{apr}(X)$, that is to say, $x \in \underline{apr}_P(\overline{apr}_P(X))$, then $\underline{apr}_P(\overline{apr}_P(X)) \supseteq \overline{apr}(X)$, thus $\underline{apr}_P(\overline{apr}_P(X)) = \overline{apr}(X)$.
11. From formula (2.1), we can get $\underline{apr}_P(\underline{apr}_P(X)) \subseteq \underline{apr}_P(X)$, and let $x \in \underline{apr}(X)$, then $I(x) \subseteq X$, thus $\underline{apr}(I(x)) \subseteq \underline{apr}(X)$. For $\underline{apr}(I(x)) \subseteq I(x)$, then $I(x) \subseteq \underline{apr}(X)$, thus $x \in \underline{apr}_P(\underline{apr}_P(X))$. According to formula (2.1), we can get $\underline{apr}_P(X) \subseteq \underline{apr}_P(\underline{apr}_P(X))$, thus $\underline{apr}_P(\underline{apr}_P(X)) = \underline{apr}_P(X)$.

If $\underline{apr}(X) = \overline{apr}(X)$, then set X can be defined in S; otherwise, set X cannot be defined in S. There are four types of set that cannot be defined on universe U, then we suppose that X is a set that cannot be defined in S, then

1. If $\underline{apr}(X) \neq \phi$ and $\overline{apr}(X) \neq U$, then the set X can be roughly defined in S.
2. If $\underline{apr}(X) \neq \phi$ and $\overline{apr}(X) = U$, then the set X cannot be defined outside in S.
3. If $\underline{apr}(X) = \phi$ and $\overline{apr}(X) \neq U$, then the set X cannot be defined inside in S.
4. If $\underline{apr}(X) = \phi$ and $\overline{apr}(X) = U$, then the set X cannot be defined totally in S.

The intuitionistic meaning of the above definitions is

If the set X in S can be roughly defined, it means that we can use some approximations to define the set X, for example, its lower approximation and upper approximation can be defined in S; if the set X cannot be defined outside S, it means that we cannot exclude that any objects in X belong to U; if the set X cannot be defined in S, it means that we cannot be sure of the fact that any object $x \in U$ belong to X; if the set X totally cannot be defined, that is to say, $\underline{apr}(X) = \phi$ and $\overline{apr}(X) = U$, then it means that we cannot even define its approximation. In such a case, these two approximations are valueless. The meaning of these definitions is obvious: it simply explains how many un-depicted levels exist between the depicted and the un-depicted. In other words, if some attributes of an object are given and we want to depict the subset of an object through these attributes, it may be failed, for only the discernible set can be uniquely depicted with the given attribute set.

Example 2.1 Suppose that the system given is $S = (U, A, V, f)$, where $U = \{x_1, x_2, \ldots, x_{10}\}$, $A = \{a_1, a_2, a_3\}$, $V_{a_1} = \{0, 1, 3\}$, $V_{a_2} = \{0, 1\}$, $V_{a_3} = \{0, 1, 2, 3\}$, furthermore, the given information function is shown in Table 2.1.

1. Please calculate the equivalence class of the information system.
2. Suppose $Y_1 = \{x_1, x_2, x_3, x_7, x_{10}\}$, $Y_2 = \{x_2, x_3, x_4, x_5, x_6, x_7, x_8\}$, $Y_3 = \{x_1, x_2, x_3, x_7, x_8\}$, $Y_4 = \{x_1, x_3, x_9\}$, please estimate whether Y_1, Y_2, Y_3, and Y_4 can be discernible in S.

Table 2.1 Information System

U	a_1	a_2	a_3
x_1	0	1	2
x_2	1	0	0
x_3	1	0	0
x_4	0	0	1
x_5	0	0	1
x_6	3	1	0
x_7	1	0	0
x_8	3	1	0
x_9	3	1	3
x_{10}	0	1	2

Solution

1. There are the following equivalence classes shown in Table 2.2 in the information system.

$$X_1 = \{x_1, x_{10}\}, \quad X_2 = \{x_2, x_3, x_7\}, \quad X_3 = \{x_4, x_5\},$$
$$X_4 = \{x_6, x_8\}, \quad X_5 = \{x_9\}$$

2. That $Y_1 = \{x_1, x_2, x_3, x_7, x_{10}\} = X_1 \cup X_2$ and $Y_2 = \{x_2, x_3, x_4, x_5, x_6, x_7, x_8\} = X_2 \cup X_3 \cup X_4$ in S are discernible, while $Y_3 = \{x_1, x_2, x_3, x_7, x_8\}$ and $Y_4 = \{x_1, x_3, x_9\}$ in S are not discernible.

Table 2.2 Equivalence Class in Information System

U		a_1	a_2	a_3
X_1	$\{x_1, x_{10}\}$	0	1	2
X_2	$\{x_2, x_3, x_7\}$	1	0	0
X_3	$\{x_4, x_5\}$	0	0	1
X_4	$\{x_6, x_8\}$	3	1	0
X_5	$\{x_9\}$	3	1	3

Example 2.2 In Example 2.1, suppose $Y_1 = \{x_1, x_2, x_4, x_5\}$, $Y_2 = \{x_1, x_2, x_3, x_4, x_6, x_9\}$, $Y_3 = \{x_1, x_2, x_5, x_8\}$, please judge whether Y_1, Y_2, Y_3 are discernible in S, and determine if they are not discernible, which kind of indiscernible set they belong to?

Solution: Y_1, Y_2, and Y_3 are indiscernible in S, among which the set $Y_1 = \{x_1, x_2, x_4, x_5\}$ is roughly defined in S; the set $Y_2 = \{x_1, x_2, x_3, x_4, x_6, x_9\}$ can be defined outside S; and the set $Y_3 = \{x_1, x_2, x_5, x_8\}$ cannot be defined inside S.

2.1.3 Attributes Dependence and Approximation Accuracy

According to indiscernible relationship, it is convenient to define some important characteristics of the information system, among which the most important characteristic is the dependency of attributes. If the number of equivalent types (element sets) derived from attribute set A is the same as that derived from $A - a_i$, then attribute a_i is regarded as redundant; otherwise, the attribute a_i is indispensable in A.

Definition 2.4 Let an information system $S = (U, A, V, f)$, and I is an indiscernible relation in U, if $P \subseteq A$, $Q \subseteq A$, then the dependency of attribute set can be given as follows:

1. If $I(P) \subseteq I(Q)$, then the attribute set Q is depended on the attribute set P, marked as $P \Rightarrow Q$.
2. If $P \Rightarrow Q$ and $Q \Rightarrow P$, then the attribute set P is equivalent to the attribute set Q, marked as $P \Leftrightarrow Q$.
3. If $P \Rightarrow Q$ and $Q \Rightarrow P$ both are not tenable, then the attribute set P and the attribute set Q are independent.

Table 2.3 Information System

U	a_1	a_2	a_3	a_4
x_1	1	1	1	1
x_2	1	0	1	2
x_3	0	0	1	0
x_4	0	0	1	0
x_5	1	0	0	2

Example 2.3 Let an information system $S = (U, A, V, f)$, among which $U = \{x_1, x_2, x_3, x_4, x_5\}$, $A = \{a_1, a_2, a_3, a_4\}$, $V_{a_1} = \{0,1\}$, $V_{a_2} = \{0,1\}$, $V_{a_3} = \{0,1\}$, $V_{a_4} = \{0,1,2\}$, as shown in Table 2.3.

In this information system the attribute set is dependent, because the number of element sets of the newly acquired information system by removing the attribute a_4 and repetitive lines and the original information system is the same, as shown in Table 2.4.

Table 2.4 Information System after Removing the Attribute a_4 and the Repetitive Lines

U	a_1	a_2	a_3
x_1	1	1	1
x_2	1	0	1
x_3	0	0	1
x_5	1	0	0

Thus the attribute a_4 is redundant. After moving a_3 from Table 2.4, the acquired information system is shown in Table 2.5.

We can see from Table 2.5, the second line and the fourth line are the same, which means that the second and fourth elements are indiscernible, that is to say, the number of discernible element sets is smaller than that in the whole attribute set; thus, the attribute a_3 is not redundant. Similarly, only three elements can be distinguished after either a_1 or a_2 is removed, which means that the number of element set is reduced; thus a_1 and a_2 are indispensable, and $\{a_1, a_2, a_3\}$ is a dispensable attribute set that has the same number of element sets as the original attribute set. Through this method, we can generate all the independent attribute sets of the attribute set a_1, a_2, a_3, a_4:$\{a_1, a_2, a_3\}$ and $\{a_3, a_4\}$.

Definition 2.5 The given knowledge representation system $S = (U, A, V, f)$, $P \subseteq A$ and $X \subseteq U$, then the definition of approximate accuracy of X is

$$\alpha_P(X) = \frac{|\underline{apr}_P(X)|}{|\overline{apr}_P(X)|} \tag{2.5}$$

where $|.|$ denotes the cardinality of a set, which is the number of elements involved in the set. The

Table 2.5 Information System after Removing a_3

U	a_1	a_2
x_1	1	1
x_2	1	0
x_3	0	0
x_5	1	0

approximate accuracy reflects comprehension degree for X according to the existing knowledge. Obviously, $0 \leq \alpha_P(X) \leq 1$, if $\alpha_P(X) = 1$ then the set X is an exact set with respect to I; if $\alpha_P(X) < 1$, then the set X is rough set with respect to I.

Example 2.4 Please solve the approximate accuracy of the set Y_1 in Example 2.2.

Solution: The base of lower approximation of the set Y_1 in Example 2.2 is 2, the base of upper approximation is 7, so the approximate accuracy of the set Y is $\alpha_P(Y_1) = |\underline{apr}_P(Y_1)|/|\overline{apr}_P(Y_1)| = 2/7$, that is to say, the set Y can be defined by the lower approximation and upper approximation in U.

2.1.4 Quality of Approximation and Reduct

To measure the dependency of knowledge, the classification quality must be defined.

Definition 2.6 Suppose $X = \{X_1, X_2, \ldots, X_n\}$ is a partition of universe U, where $X_i(i = 1, 2, \ldots, n)$ is one class of X, and $P \subseteq A$, then the quality of approximation of X is

$$\gamma_P(X) = \frac{\sum_{i=1}^{n} |\underline{apr}_P(X_i)|}{|U|} \tag{2.6}$$

where $|.|$ denotes the cardinality of a set, and the quality of approximation $\gamma_P(X)$ represents the ratio of all correctly classified objects by the attribute set P to all objects in the system.

If the quality of approximation $\gamma_P(X) = 1$, then the knowledge X is completely dependent on P: if $0 < \gamma_P(X) < 1$, then we can say that the knowledge X is partly dependent on P, which reveals that only partial attributes in the P are available, or the dataset has some defects from the beginning. What's more, the complementary nature of $\gamma_P(X)$ gives a contradictory measurement in the chosen data subsets. If $\gamma_P(X) = 0$, then the knowledge X is completely independent on P.

When we remove some attribute from conditional attribute set appointed, it is available to define the importance of an attribute through calculating the change of dependency. Suppose $Q \subseteq A$, $p \in P$, and $P \subseteq A$, then the attribute importance $sgf(p, Q)$ is

$$sgf(p,Q) = \gamma_P(Q) - \gamma_{P-\{p\}}(Q) \tag{2.7}$$

The more the change of the dependence is, the more important P is; thus the choice of attribute refers to the process of ruling out those attributes that are not very important for the present mission of pattern classification.

If the attribute set is dependent, then it is available to search for the smallest attribute subset that has the same number of element sets compared with the whole attribute set. To check whether the attribute set is independent, it is necessary to check whether it will reduce the number of element sets after removing every attribute. Core and attribute reduct are two fundamental concepts while reduct is the smallest independent attribute subset that has the same data division with the overall attribute sets, and it is the essential part of the information system, which can be used to distinguish all the objects that can be discernible in the original information system. The core is the common part of all reducts.

2.1.5 Calculation of the Reduct and Core of Information System Based on Discernable Matrix

Showeron put forward a method to express knowledge with the discernable matrixes in 1991, which has many advantages. In particular, it can conveniently explain and calculate the core and reduct of the information system.

Definition 2.7 Suppose that $S = (U, A, V, f)$ is an information system, $|U| = n$ and the discernable matrix of S is $n \times n$, among which any of its elements is $d(x, y) = \{a \in A | f(x, a) \neq f(y, a)\}$; thus, $d(x, y)$ is an attribute set that can discern the object x and y.

Definition 2.8 For every attribute $a \in A$, appoint a Boolean variable "a," if $a(x, y) = \{a_1, a_2, \ldots, a_k\} \neq \phi$, Boolean function is $a_1 \vee a_2 \vee \ldots a_k$, which is marked as $\sum a(x, y)$; if $a(x, y) = \phi$, and Boolean constant is 1, then the discernibility function is defined as

$$f(A) = \prod_{(x,y) \in U \times U} \sum a(x, y)$$

If the Boolean expression can only be represented by Boolean variables and constants through the operation of disjunction and conjunction, then the Boolean expression is called a paradigm; if Boolean expression is a paradigm formed by the conjunction that is composed of some disjunctive forms, then the Boolean expression is named a conjunctive paradigm; if the Boolean expression is a paradigm consisting of disjunctions that is composed of some conjunctive forms, then the Boolean expression is called a disjunctive paradigm; if the Boolean expression is a disjunctive paradigm and involves the smallest number of conjunctive forms, then the Boolean expression is called a minimal disjunctive paradigm.

The properties of the discernibility function $f(A)$ are as follows: All the conjunctive forms in the minimal disjunctive paradigm of discernibility function $f(A)$ are the whole reducts of attribute set A.

The manifestation of discernable matrix is an $n \times n$ dimensional matrix, where n stands for the number of elements, and its elements are defined as the set that involves all the attributes of the discernibility element set $[x]_i$ and $[x]_j$, and represented as d_{ij}. Thus, to calculate element d_{ij}, the attributes of the element set $[x]_i$ and $[x]_j$ should be recognized.

Obviously, $d_{ij} = d_{ji}$, $d_{ij} \neq \phi$, and discernable matrix is symmetrical, thus the discernable matrix can be expressed only by calculating the lower triangular matrix. Calculating according to operation principles such as absorptivity and distribution principles for discernable function of the information system, we can obtain the minimal disjunctive paradigm of discernibility function, that is to say, we can obtain the reduct and core of the information system.

Example 2.5 As for the discernable matrix D of the five element sets shown in Table 2.1, the equivalence classes in the system are as follows: $X_1 = \{x_1, x_{10}\}$, $X_2 = \{x_2, x_3, x_7\}$, $X_3 = \{x_4, x_5\}$, $X_4 = \{x_6, x_8\}$, $X_5 = \{x_9\}$. The construction of discernable matrix is as follows.

The attribute set of recognition element set X_1 and X_2 includes the attribute $\{a_1, a_2, a_3\}$, and the attribute a_2, a_3 have only recognized the element set X_3 and X_1, thus the elements $d_{31} = d_{13} = \{a_2, a_3\}$. Because the discernable matrix is symmetric ($d_{ij} = d_{ji}$), we only need to consider the lower triangular part. To make the discernable matrix no null, every element set should have at least one attribute that is different from other element sets. The discernable matrix constructed by the above five element sets is shown in Table 2.6.

A discernable matrix can be used to search for the minimal attribute subset (reduct) to derive the same data distraction as in attribute set A. To find the minimal attribute subset, it is necessary to construct a discernibility function that is a Boolean function and can be constructed in the following method.

For each attribute that can identify two element sets, such as a_1, a_2, a_3, appoint a Boolean constant a, the form of the Boolean function is $a_1 + a_2 + a_3$ or $(a_1 \vee a_2 \vee a_3)$. If the attribute set is empty, then Boolean constant is 1. For example, in regard to the discernable matrix showed in Table 2.6, the discernibility function is

$$f(A) = (a_1 + a_2 + a_3)(a_2 + a_3)(a_1 + a_3)(a_1 + a_3)$$
$$\times (a_1 + a_3)(a_1 + a_2)(a_1 + a_2 + a_3) \times (a_1 + a_2 + a_3)(a_1 + a_2 + a_3)a_3$$

Table 2.6 Discernable Matrix

	Set 1	*Set 2*	*Set 3*	*Set 4*	*Set 5*
Set 1					
Set 2	a_1, a_2, a_3				
Set 3	a_2, a_3	a_1, a_3			
Set 4	a_1, a_3	a_1, a_2	a_1, a_2, a_3		
Set 5	a_1, a_3	a_1, a_2, a_3	a_1, a_2, a_3	a_3	

To calculate the final form of $f(A)$, we should use the absorptivity law, if the attributes a_1, a_2, a_3 can discern the element set X_1 and X_2, and the attributes a_2, a_3 can discern X_3, then we can only consider the attribute a_2, a_3, because the attribute a_2, a_3 can discern the set X_1, X_2, and X_3. For example,

$$(a_1 + a_2 + a_3)(a_2 + a_3) = a_2 + a_3$$

Another example, from the first line in Table 2.6, is that if we want to discern the element set X_1 from the element set X_2, X_3, X_2, X_3, X_4, and X_5, we should consider the following attribute sets: $\{a_1, a_2, a_3\}$, $\{a_2, a_3\}$, $\{a_1, a_3\}$, and $\{a_1, a_3\}$. Then the discernibility function is

$$(a_1 + a_2 + a_3)(a_2 + a_3)(a_1 + a_3)(a_1 + a_3) = a_2a_1 + a_2a_3 + a_3a_1 + a_3$$

In fact, it is simple to calculate the core and reduct through discernable matrixes. The core is a set of all the individual elements that are in discernable matrix, while reduct is the minimal attribute subset, and it has at least one common element with any nonempty item of the discernable matrix.

As shown in Table 2.6, we can obtain

$$\begin{aligned} f(A) &= (a_1 + a_2 + a_3)(a_2 + a_3)(a_1 + a_3)(a_1 + a_3) \\ &\quad \times (a_1 + a_3)(a_1 + a_2)(a_1 + a_2 + a_3) \\ &\quad \times (a_1 + a_2 + a_3)(a_1 + a_2 + a_3)a_3 \\ &= (a_2 + a_3)(a_1 + a_3)(a_1 + a_2)a_3 \\ &= a_3(a_1a_2 + a_1a_3 + a_2a_3 + a_3)(a_1 + a_2) \\ &= a_3(a_1a_2 + a_1a_3 + a_1a_2a_3 + a_1a_3 + a_1a_2 + a_1a_2a_3 + a_2a_3 + a_2a_3) \\ &= a_3(a_1a_2 + a_1a_3 + a_2a_3) \end{aligned}$$

Based on this method, we find two reducts of the attribute set $\{a_2, a_3\}$ and $\{a_1, a_3\}$. Generally, there are many reducts in an information system, while we can use one of them to express the information system, such as $\{a_1, a_3\}$; the information system expressed is shown in Table 2.7.

The expression of knowledge in Table 2.7 is equal to the form in Table 2.2, while reduct $\{a_1, a_3\}$ and $\{a_2, a_3\}$ are similar to the attribute set $\{a_1, a_2, a_3\}$. It provides the same knowledge partition about universe U.

Core appears to be all single attribute sets in the discernable matrix. In this example, the core is a_3.

Through the reduction of some unnecessary attribute sets, the information system can be simplified. For example, by excluding some attribute values, we can identify all the elementary sets in the system.

Table 2.7 Information System Generated by Reduct $\{a_1, a_3\}$

U	a_1	a_3
$\{x_1, x_{10}\}$	0	2
$\{x_2, x_3, x_7\}$	1	0
$\{x_4, x_5\}$	0	1
$\{x_6, x_8\}$	3	0
$\{x_9\}$	3	3

Table 2.8 Discernibility Function Constructed by Reduct $\{a_1, a_3\}$

	Set 1	Set 2	Set 3	Set 4	Set 5
Set 1		a_1, a_3	a_3	a_1, a_3	a_1, a_3
Set 2	a_1, a_3		a_1, a_3	a_1	a_1, a_3
Set 3	a_3	a_1, a_3		a_1, a_3	a_1, a_3
Set 4	a_1, a_3	a_1	a_1, a_3		a_3
Set 5	a_1, a_3	a_1, a_3	a_1, a_3	a_3	

The reduction process of attribute value is similar to the searching of the core and the reducting of attribute value. It can also be finished through discernable matrix, but the definition of discernibility function is a little different. The reduction of attribute value needs to construct many discernibility functions instead of one discernibility function.

Example 2.6 Make reduction of attribute value about the information system Table 2.7 generated by the reduct $\{a_1, a_2\}$ in Example 2.5.

Based on the five element sets in reduct $\{a_1, a_2\}$ and Table 2.7, we can construct five discernibility functions: $f_1(A)$, $f_2(A)$,…, $f_5(A)$, where discernibility function $f_1(A)$ is the attribute set that can distinguish the elementary set 1 and the other four elementary sets 2, 3, 4, 5. Discernibility function $f_2(A)$ is the attribute set distinguished between elementary set 2 and the elementary sets 1, 3, 4, 5. Similarly, we can construct the discernibility functions shown in Table 2.8.

To construct $f_1(A)$, we must consider all the attribute sets in the first column of Table 2.8; while to construct $f_2(A)$, we must consider all the attribute sets in the second column of Table 2.8. Similarly,

$$f_1(A) = (a_1 + a_3)a_3(a_1 + a_3)(a_1 + a_3) = a_3$$

$$f_2(A) = (a_1 + a_3)(a_1 + a_3)a_1(a_1 + a_3) = a_1$$

$$f_3(A) = (a_1 + a_3)(a_1 + a_3)a_3(a_1 + a_3) = a_3$$

$$f_4(A) = (a_1 + a_3)a_1a_3(a_1 + a_3) = a_1a_3$$

$$f_5(A) = (a_1 + a_3)(a_1 + a_3)(a_1 + a_3)a_3 = a_3$$

According to $f_1(A)$, we can draw a conclusion without considering a_1; according to $f_2(A)$, we can draw a conclusion without considering a_3; according to $f_3(A)$, we can draw a conclusion without considering a_1; according to $f_4(A)$, the attribute values a_1 and a_3 have to be considered, because the attribute value has a reduct $\{a_1, a_3\}$; according to $f_5(A)$, we can draw a conclusion without considering a_1. So, the form of attribute value reduct of the reduct $\{a_1, a_3\}$ is shown in Table 2.9

Table 2.9 Table of Attribute Value Reduct

U	a_1	a_3
$\{x_1, x_{10}\}$	*	2
$\{x_2, x_3, x_7\}$	1	*
$\{x_4, x_5\}$	*	1
$\{x_6, x_8\}$	3	0
$\{x_9\}$	*	3

Note: * represents those that "need not be considered."

Example 2.7 Classify the following object sets by using the data in Table 2.1:

$$Y_1 = \{x_2, x_3, x_7\},\quad Y_2 = \{x_1, x_6, x_8\},\quad Y_3 = \{x_4, x_5, x_1, x_{10}, x_2\}$$

Solution: Equivalence classes generated by Table 2.1 are

$$X_1 = \{x_1, x_{10}\},\quad X_2 = \{x_2, x_3, x_7\},\quad X_3 = \{x_4, x_5\},\quad X_4 = \{x_6, x_8\},\quad X_5 = \{x_9\}$$

The lower approximations, the upper approximations, and the accuracy of classification of every object set are shown in Table 2.10.

So the total classification accuracy is

$$\alpha_P(X) = \frac{|\underline{apr}(X)|}{|\overline{apr}(X)|} = \frac{3+2+4}{3+4+7} = \frac{9}{14} = 64.3 \text{ percent}$$

The total quality of classification is

$$\gamma_P(X) = \frac{\sum_{i=1}^{n} |\underline{apr}(X_i)|}{|U|} = \frac{9}{11} = 81.8 \text{ percent}$$

When removing an attribute from the reduct, we can also calculate the classification accuracy of every class. Taking the reduct $\{a_2, a_3\}$ as an example, when we remove the attribute a_2 from the reduct $\{a_2, a_3\}$, the classification accuracy of the classes Y_1 and Y_2 both experience dramatic changes and are both internally indefinable. Meanwhile, the classification accuracy of the category Y_3 also changes. When we remove the attribute a_3, all the categories are indefinable: the categories Y_1 and Y_2 are internally indefinable and the category Y_3 is completely indefinable. Table 2.11 reveals how the attribute influences the classification accuracy of the equivalence class describing the object sets.

Table 2.10 Lower Approximations, Upper Approximations, and Classification Accuracies of Every Object Set

Class	*Object Numbers*	*Lower Approximations*	*Upper Approximations*	*Classification Accuracies*
Y_1	3	3	3	1
Y_2	3	2	4	0.5
Y_3	5	4	7	0.57

Table 2.11 Classification Accuracy after Removing Attributes from the Reduct {a_2, a_3}

	Move Attribute		
Category	*Do Not Move*	*Move a_2*	*Move a_3*
Y_1	1	Internally indefinable	Internally indefinable
Y_2	0.5	Internally indefinable	Internally indefinable
Y_3	0.57	0.44	Completely indefinable

2.2 Decision Table and Rule Acquisition

The concept of attribute dependence can be used to describe the relationship between causes and results in dataset. According to date reduction the identified and possible decision rules can be achieved by absolute dependence and partial dependence.

2.2.1 The Attribute Dependence, Attribute Reduct, and Core

Definition 2.9 If in an information system $S = (U, A, V, f)$, the attributes of A can be divided into two disjoint subsets, which are condition attributes C and decision attributes D, and that is $A = C \cup D$, $C \cap D = \phi$, then S is called decision table.

Dependence of attributes is a very important concept in the practical application of decision table. Intuitively, if all values of the attributes from D are uniquely determined by the values of attributes from C, that is, D completely depends on C, there exists functional dependence in the values between C and D; if only some values of attributes from D are determined by the values of attributes from C, then it is called partial functional dependence. To a certain extent, functional dependence reveals the degree that construction of related knowledge granularity of C can represent the construction of the knowledge granularity of D. Different grade of information generalization can have different dependences, and strong function in high grade knowledge granularity dependence implies low function dependence. In rough set approach, discovering dependence is very important for knowledge analysis, data mining, and general data reasoning.

The classification quality of reduct attribute sets is the same as that of the original attribute sets. If the minimal attribute set $P \subseteq C \subseteq A$ satisfies $\gamma_P(X) = \gamma_C(X)$, then set P is called a reduct of C, and is denoted as $RED(P)$. Simply speaking, reduct is the minimal subset of the condition attributes without redundant attributes and it guarantees right classification. The discovery of attribute dependence results in the generation of the minimal attribute subset reduct that has the same quality of classification as that of the original attribute set.

More than one reduct may exist in the decision table, and the intersection of all reducts is called the core of decision table, which is denoted as

$$CORE(P) = \bigcap_{R_i \in RED(P)} R_i, \quad (i = 1, 2, \ldots, n) \tag{2.8}$$

Core is the most important set of attributes in the information system, and it may be an empty set.

2.2.2 Decision Rules

In any learning system, rule generation is a very important task. The set of all condition elements in the universe is called the condition classes of S, denoted by $X_i (i = 1, 2, \ldots, n)$. The set of all decision elements in the universe is called the decision classes of S, denoted by $Y_j (j = 1, 2, \ldots, n)$, if $X_i \cap Y_j = \phi$, then

$$r : Des_C(X_i) \Rightarrow Des_D(Y_j) \tag{2.9}$$

It is called the decision rules of (C, D), denoted as $\{r_{ij}\}$. For $\forall i, j$, if $X_i \subseteq Y_j$, then rule r_{ij} is decisive in S, otherwise it is indecisive. The syntax of a rule is as follows:

If $f(x, q_1) = r_{q1} \wedge f(x, q_2) = r_{q2} \wedge \cdots \wedge f(x, q_p) = r_{qp}$, then $x \in Y_{j1} \vee Y_{j2} \vee \cdots \vee Y_{jk}$, where $\{q_1, q_2, \ldots, q_p\} \subseteq C$; $(r_{q_1}, r_{q_2}, \ldots, r_{q_p}) \in V_{q_1} \times V_{q_2} \times \cdots V_{q_p}$.

The "if" part of decision rule is called the condition part, while the "then" part is called the decision part. If the consequences are univocal, that is, one object not only matches the condition part but also the decision part, then $k = 1$, the rule is exact, or the rule is approximate. The number of objects that support the decision rules is called the supporting number.

It is a complex task to generate decision rules from decision tables, and the solutions proposed to solve it at present are usually based on one of the following tactics:

1. Generating a minimal set of rules that covers all objects from a decision table.
2. Generating an exhaustive set of rules that consists of all possible rules from a decision table.
3. Generating a set of "strong" decision rule sets, or even a partially distinguished set, within which every rule covers relatively more objects but not necessarily all the objects from a decision table.

2.2.3 Use the Discernibility Matrix to Work Out Reducts, Core, and Decision Rules of Decision Table

The main steps using discernibility matrix to calculate the reducts and core of a decision table are

1. Construct the sets of elements based on condition attribute sets and decision attribute sets respectively.
2. Calculate upper approximation and lower approximation of decision attribute sets.
3. Search for the reducts and core from decision table.
4. Reduce the decision rule sets.

Example 2.8 A decision table containing three sets of condition attributes $\{a_1, a_2, a_3\}$ and a set of decision attributes d is shown in Table 2.12. Calculate the reducts and core from the decision table.

Solution: Based on the decision attributes, we divide universe and obtain the following equivalence classes:

$$\frac{U}{D} = \{\{x_1, x_3, x_9\}, \{x_2, x_4, x_7, x_{10}\}, \{x_5, x_6, x_8\}\}$$

$$\frac{U}{C} = \{\{x_1, x_3, x_9\}, \{x_2, x_7, x_{10}\}, \{x_4\}, \{x_5, x_8\}, \{x_6\}\}$$

Table 2.12 Decision Table

U	a_1	a_2	a_3	d
x_1	1	2	4	2
x_2	4	1	2	1
x_3	1	2	4	2
x_4	1	1	4	1
x_5	2	2	3	4
x_6	2	2	1	4
x_7	4	1	2	1
x_8	2	2	3	4
x_9	1	2	4	2
x_{10}	4	1	2	1

Table 2.13 Upper Approximation, Lower Approximation, and Approximate Quality of Decision Table

Decision Class	*Support Object Number*	*Upper Approximation*	*Lower Approximation*	*Approximate Quality*
$\{x_1, x_3, x_9\}$	3	3	3	1
$\{x_2, x_4, x_7, x_{10}\}$	4	4	4	1
$\{x_5, x_6, x_8\}$	3	3	3	1

According to the definitions in Section 2.1, the upper approximation, lower approximation, and approximate quality of decision table can be obtained and shown in Table 2.13. We can see from Table 2.13 that approximate quality of decision class is 1. That is to say, based on condition attribute set $\{a_1, a_2, a_3\}$, the sets of elements in the decision attribute set $\{d\}$ can be all described correctly.

To calculate the reducts and core from decision table with a discernibility matrix, first, we need to construct a discernibility matrix that can distinguish objects of different classes in the $\{d\}$ (the elements in the discernibility matrix distinguish objects of different classes in the $\{d\}$).

In decision classes, because the object x_1 and x_3, x_9 appear in the same class, it need not be distinguished. As a result of this, lines 1, 3, and 9 of the first column in Table 2.12 need not be considered, and lines 2, 4, 7, and 10 of the second column in Table 2.12 need not be considered, and the relative lines of the other columns are not considered accordingly. The discernibility function constructed is shown in Table 2.14.

Based on data in Table 2.14, the discernibility function $f_4(D)$ is obtained:

Table 2.14 Discernibility Function Constructed by Decision Table 2.12

	1	2	3	4	5	6	7	8	9
2	a_1, a_2, a_3								
3	—	a_1, a_2, a_3							
4	a_2	—	a_2						
5	a_1, a_3	a_1, a_2, a_3	a_1, a_3	a_1, a_2, a_3					
6	a_1, a_3	a_1, a_2, a_3	a_1, a_3	a_1, a_2, a_3	—				
7	a_1, a_2, a_3	—	a_1, a_2, a_3	—	a_1, a_2, a_3	a_1, a_2, a_3			
8	a_1, a_3	a_1, a_2, a_3	a_1, a_3	a_1, a_2, a_3	—	—	a_1, a_2, a_3		
9	—	a_1, a_2, a_3	—	a_2	a_1, a_3	a_1, a_3	a_1, a_2, a_3	a_1, a_3	
10	a_1, a_2, a_3	—	a_1, a_2, a_3	—	a_1, a_2, a_3	a_1, a_2, a_3	—	a_1, a_2, a_3	a_1, a_2, a_3

$$
\begin{aligned}
f_4(D) = {} & (a_1 + a_2 + a_3)a_2(a_1 + a_3)(a_1 + a_3) \\
& \times(a_1 + a_2 + a_3)(a_1 + a_3)(a_1 + a_2 + a_3) \\
& \times(a_1 + a_2 + a_3)(a_1 + a_2 + a_3) \\
& \times(a_1 + a_2 + a_3)(a_1 + a_2 + a_3) \\
& \times(a_1 + a_2 + a_3)a_2(a_1 + a_3)(a_1 + a_3) \\
& \times(a_1 + a_2 + a_3)(a_1 + a_3)(a_1 + a_2 + a_3) \\
& \times(a_1 + a_2 + a_3)(a_1 + a_2 + a_3) \\
& \times(a_1 + a_2 + a_3)a_2 \\
& \times(a_1 + a_2 + a_3)(a_1 + a_3)(a_1 + a_2 + a_3) \\
& \times(a_1 + a_2 + a_3)(a_1 + a_3)(a_1 + a_2 + a_3) \\
& \times(a_1 + a_2 + a_3)(a_1 + a_2 + a_3)(a_1 + a_3) \\
& \times(a_1 + a_2 + a_3)(a_1 + a_2 + a_3) \\
= {} & a_2(a_1 + a_3) = a_1a_2 + a_2a_3
\end{aligned}
$$

Table 2.15 Decision Table Expressed by the Reduct {a_1, a_2}

U	a_1	a_2	d
x_1	1	2	2
x_2	4	1	1
x_3	1	2	2
x_4	1	1	1
x_5	2	2	4
x_6	2	2	4
x_7	4	1	1
x_8	2	2	4
x_9	1	2	2
x_{10}	4	1	1

Obviously, $\{a_1, a_2\}$ and $\{a_2, a_3\}$ are two reducts of decision table and the core is $\{a_2\}$ (Table 2.15). That is to say, a decision table (Table 2.13) can be expressed by any form of decision tables (Table 2.12 or 2.16).

Decision rule set generated by the reduct $\{a_1, a_2\}$ is shown in Table 2.17. Similarly, decision rule set can be generated by the reduct $\{a_2, a_3\}$.

To get reduction of the unnecessary condition attribute value in decision rule set in Table 2.17, the discernable matrix constructed by the reduct $\{a_1, a_2\}$ is shown in Table 2.18.

From Table 2.18, relative discernibility functions are obtained:

$$f_1(D) = (a_1 + a_2)a_2a_1 = a_1a_2$$

$$f_2(D) = (a_1 + a_2)(a_1 + a_2) = a_1 + a_2$$

$$f_4(D) = a_2(a_1 + a_2) = a_2$$

$$f_5(D) = a_1(a_1 + a_2)(a_1 + a_2) = a_1$$

Table 2.16 Decision Table Represented by the Reduct {a_2, a_3}

U	a_2	a_3	$d1$
x_1	2	4	2
x_2	1	2	1
x_3	2	4	2
x_4	1	4	1
x_5	2	3	4
x_6	2	1	4
x_7	1	2	1
x_8	2	3	4
x_9	2	4	2
x_{10}	1	2	1

Table 2.17 Rule Set Generated by the Reduct $\{a_1, a_2\}$

Decision Rules	*Supporting Number*
If $a_1 = 1$ and $a_2 = 2$ then $d = 2$	3
If $a_1 = 4$ and $a_2 = 1$ then $d = 1$	3
If $a_1 = 1$ and $a_2 = 1$ then $d = 1$	1
If $a_1 = 2$ and $a_2 = 2$ then $d = 4$	3

Table 2.18 Discernable Matrix Constructed by the Reduct $\{a_1, a_2\}$

	1	2	4	5
1	—	a_1, a_2	a_2	a_1
2	a_1, a_2	—	—	a_1, a_2
4	a_2	—	—	a_1, a_2
5	a_1	a_1, a_2	a_1, a_2	—

Table 2.19 Minimal Decision Rule Set Generated by the Reduct $\{a_1, a_2\}$

Decision Rules	*Supporting Number*
If $a_1 = 1$ and $a_2 = 2$ then $d = 2$	3
If $a_1 = 4$ or $a_2 = 1$ then $d = 1$	3
If $a_2 = 1$ then $d = 1$	1
If $a_1 = 2$ then $d = 4$	3

Hereby, the minimal decision rule set generated by the reduct $\{a_1, a_2\}$ is shown in Table 2.19.

2.3 Data Discretization

Rough set theory analysis requires that data appear in the form of categories. However, most data are digital data in actual application. Because rough set theory does not include digital data discretization, we must make use of a proper way to convert the digital data into discrete interval before the application of the rough set method, that is, we need to transform the continuous number into a series of natural numbers. Continuous data discretization may result in a decrease of the precision of original data, but improved the generality. Therefore, digital data discretization can be regarded as a pretreatment of rough set method.

Data discretization can be divided into expert discretization and automatic discretization. The latter means to discretize data in an automatically defined way. Automatic discretization method can also be divided into supervised discrete method and unsupervised discrete method. The latter can be regarded as a simple clustering process, such as equal frequency interval method and equal width interval method. The supervised discretization method can be divided into globe discrete method and local discrete method according to the discrete method applied in the current set value or in all the values in the universe.

Globe discrete method must take into consideration condition attribute groups during the discretization. First of all, the objects are divided into several categories according to the similarity between all the condition attributes (minimum European distance); then we should determine whether adjacent intervals are to be combined through checking consistency level of every attribute. If consistency level of combined intervals is equal to or higher than that of the original data, we can combine these intervals, and the process continues until there is no consistency to be improved. If discrete results contain nearly the same consistency level as that of the original data, then the discrete performance and the reduction from the results are optimal. In the globe discrete method, we should put all the values of the universe into the discretizational set and, generally speaking, increasing or amending the value in the set will not affect the discrete results.

The local discrete method, which includes the minimum entropy method as well as the Chimerge method (chi square statistical variables taken as a fitness measurement), is applied only in the subregion of the decision table and to the discretization of the values set, and not to those of other sets. Local discrete method often contains a discretization of continuous condition attribute variable at one time, and increasing or amending set values may lead to rediscrete dataset.

Data discrete problem is a complete NP problem; however, how discrete attribute values can be optimized has not yet solved, and it is worth doing further research on this subject. Common data discrete methods are as follows.

2.3.1 Expert Discrete Method

It means the discretization conducted by an expert in a particular field according to his judgment or the definite standard in the field. Generally speaking, data collected from different application fields have their own characteristics and trends. Experts from different fields are often referred to those who have the necessary knowledge and experience in the fields, and they can put forward advices on how to divide a certain range of data into some intervals according to the sizes of the numbers and intervals. For example, it is well known that the normal body temperature is 97.7°F; temperatures between 97.7°F and 102°F indicate slight fever, while temperatures between 102°F and 104°F indicate serious fever. We can tell the threshold in this case from our own knowledge and experience. Obviously, experts can offer more reasonable dividing points than an automatic discrete method, and discretization of digital data may lead to serious imprecision and decrease of the reliability of derived rules in case expert knowledge is lacking. However, not every attribute value set has an important threshold, and in some cases we have to analyze the data without necessary experience and expert knowledge, or we need to be aided by the automatic discrete method when the model contains updated indicator factors, so it is useful to study suitable discrete technique of the digital data by considering some principles and standards.

2.3.2 Equal Width Interval Method and Equal Frequency Interval Method

Equal width interval method is the simplest discrete method. This method divides the attribute value between minimum and maximum into N equal intervals (N is determined by the clients). If L and H represent, respectively, minimum and maximum, then there exists the equal width of these intervals, denoted by W, that is,

$$W = \frac{(H - L)}{N}$$

The equal frequency interval method divides the ranges of digital characteristics by choosing intervals so that the number of the training cases of each interval is approximately the same.

2.3.3 The Most Subdivision Entropy Method

It is applied especially in decision tree algorithm. This method takes class entropy as a standard to select the optimal dividing point, so that the class entropy is the minimum in the all-candidate dividing points.

2.3.4 Chimerge Method

This method is to decide whether the adjacent intervals should be merged by checking whether the classification precision is affected after two intervals are merged. For example, X_1 and X_2 are two adjacent intervals, and y_1, y_2, y_3 are three categories of a categorical variable y, and f_{ij} is data frequency between the intervals X_i classified to the category y_j, then we use chimerge method with this table to check according to the fact whether object classification of the variable y is dependent on their interval. If there exists dependence, then X_1 and X_2 should not be merged or they can be merged and will not affect the classification precision according to the variable y.

The above four methods do not consider the relationship between condition attribute and decision attribute, that is, the intervals formed only consider condition attribute but not the consistency with the decision attribute. Chimerge method and globe discrete method must take into consideration the relationship between condition attribute and decision attribute.

2.4 Common Algorithms of Attribute Reduct

Reduct is important for a series of rules constructed by the classification objects in a model. Wong and Ziarko have proved that searching for all the reducts or the best reduct of an information system is an NP-hard problem, and all of $2^{|P|}$ subsets of $\phi \subset P \subseteq C$ should be taken into consideration, whose time complexity is of index class; the reason is that the combination of attributes causes the explosion. In the field of artificial intelligence, heuristic search is a general method that can be used to solve this kind of problem, and by simplifying calculation with heuristic information we can find the best reduct or second-best reduct. In practical application, it is not necessary to calculate all the reducts, but just some of them. Generally speaking, the reduct that contains the minimal attribute number is the most satisfactory reduct. There are two research directions in the acquisition of the minimal attribute number: one is to decrease the complexity of reduct problems under the condition of guarantee of the completeness of the algorithm; the other is to find quick algorithm, and simplify calculation with heuristic information and improve efficiency of algorithm by sacrificing the completeness of algorithm, so that we can find the best reduct or second-best reduct, such as the attribute importance algorithm based on information entropy, the attribute importance algorithm based on dependence, and the attribute importance algorithm based on occurrence frequency and genetic algorithm (GA), and so on.

2.4.1 Quick Reduct Algorithm

Quick reduct algorithm can be used to find the reduct whose attribute number is the minimal. The process of quick reduct algorithm can be simply described as follows: begin with an empty set, add attributes one after another, put the attribute of the maximum classification quality $\gamma_P(X)$ into the present attribute subset every time, and repeat the process until chosen classification quality $\gamma_P(X)$

reaches 1 or equals to the whole classification quality of condition attribute. The concrete algorithm is shown as follows:

```
Quick reduct (C, D)
Input: all the condition attribute sets C and all the decision
  attribute sets D
Output: attribute reduct R ⊆ C
  R←{}
do
  T←R
    For ∀x∈(C−R)
    If γ_{R∪{x}}(D) > γ_T(D)
T←R∪{x}
R←T
    Until γ_R(D) = γ_C(D)
Return R
```

Disadvantages of the quick reduct algorithm lie in the fact that it cannot guarantee to generate an attribute set with the minimal reduct.

2.4.2 Heuristic Algorithm of Attribute Reduct

The heuristic attribute reduct method generally starts from the core attribute of the system, and the attributes are divided into the attribute set until the latter is a reduct according to the sequence of descending importance. There are some kinds of attribute importance that are often used, such as the attribute importance based on dependability, the attribute importance based on information entropy, and the attribute importance based on the property frequency in a discernable matrix, and so on.

Suppose decision table $S = (U, A, V, f)$, $A = C \cup D$, $C \cap D = \phi$, $\phi \neq P \subset C$, then the attribute importance based on dependence of the attribute $a \in C - P$ is $sgf(a, P) = g(P \cup \{a\}, D) - g(P, D)$, where $g(P, D) = |pos_P(D)|/|pos_C(D)|$.

Assume that the relative conditional entropy D to P is $H(D/P)$, and $\phi \neq P \subset C$, then the information entropy importance of attribute $a \in C - P$ is $sgf(a, P) = H(D/P) - H(D/P \cup \{a\})$.

The general procedures of heuristic attribute reduct algorithm are as follows:

Step 1: Construct the differentiation matrix and obtain the core attribute set R.

Step 2: Make the initialized attribute set $R_1 = R$.

Step 3: If the intersection of R_1 and element d_{ij} is a null set in the discernable matrix attribute, then choose a with the highest attribute importance from d_{ij}, and let $R = R \cup \{a\}$. Repeat this step to the degree that every element in the discernable matrix can intersect with R_1, i.e., $I(R_1) = I(C)$.

Step 4: Ensure that R_1 is a reduct, and subtract the excrescent attributes.

2.4.3 Genetic Algorithm

GA is a search and optimization method developed by simulating the biological heredity and long-term evolution that was first put forward by Professor Holland of Michigan University in 1975. GA simulates the mechanism of "struggle for survival, select the superior and eliminate the inferior, and survival of the fittest" to search and optimize with the step iterative method, and paves the way for overall optimization of many difficulties. GA is a community operation, its object are all the units in the group. The operator contains selection, crossover, and mutation. GA's basic flow is shown in Figure 2.2.

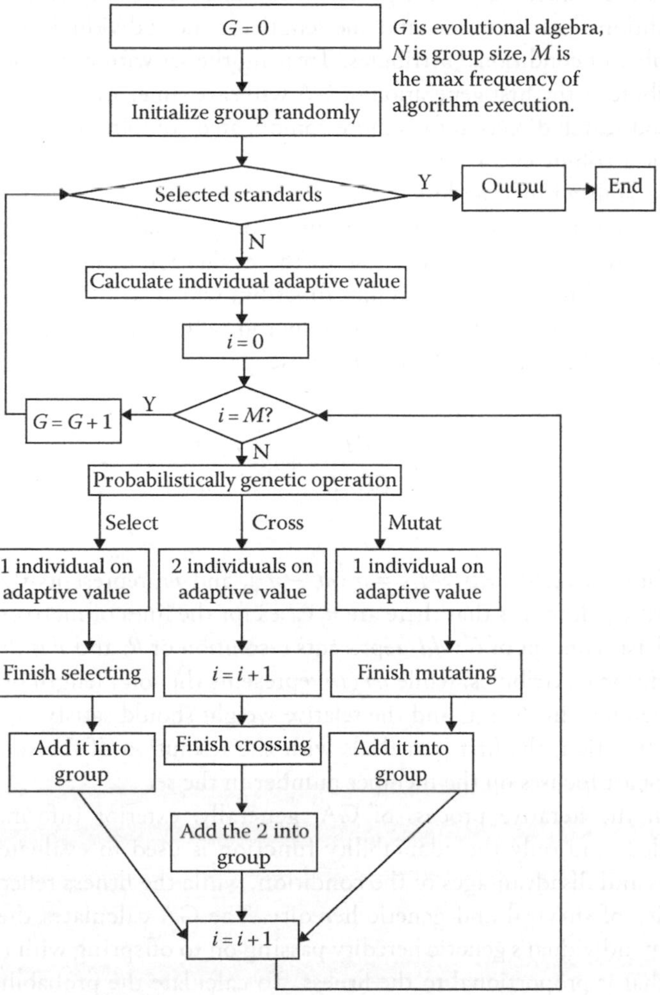

Figure 2.2 Flowchart of GA.

The key elements of GA include chromosome coding, individual adaptive evaluation, genetic operators (selection operator, crossover operator, and mutation operator), genetic parameter setting, and so on.

1. Chromosome coding
   Binary symbol strings with fixed length represent individuals of a group in GA. The allele is made up of binary value set of {0, 1}. The genetic value of every individual can be generated by evenly distributed random numbers in the initial set. For example, Y =110100101011001 is an individual, whose chromosomal length is 15.

   In the reduct search, the genetic individual of a group can be seen as the conditional attribute set, and the length of the individual depends on the number of conditional attributes. Treating the set with only one conditional attribute as the first generation of GA will save time, and the process will set a good search direction for a more complicated combination on the basis of a single attribute evaluation.
2. Adaptability function design
   This is a very important step, because a proper evaluation of the individual will lead to a benign search result. The evaluation of an individual can be achieved through an adaptability function, which will take a lot of time, so we can make use of the following adaptability function to evaluate the adaptability of the conditional attribute subset P:

$$f(P) = w_1 \frac{dd_P}{BDM_{C_i}} + w_2 \frac{n - \text{length}}{n-1} \tag{2.10}$$

In formula (2.10), $RBDM_{C_i} = n_e(m_e - n_e)$, and m_e represents the number of lines, which means that there are n_e cases for the lines of matrix contained in the basic concept of C_i; dd_P represents resolution of P; and n is the number of conditional attributes, while *length* represents the total length of the number of members in the set, and the relative weight should satisfy $w_1 + w_2 = 1$. In this function, the first part deals with the resolution of the subset, and the next part focuses on the member number in the set.

In the iterative process of GA, generally, exterior information is not needed, and only the adaptability function is used to evaluate the advantages and disadvantages of the condition, while the fitness reflects the probability of survival and genetic heredity. The GA calculates the number of every individual's genetic heredity passing on to offspring with the probability that is proportional to the fitness. To calculate the probability correctly, generally speaking, the fitness of the individual is required to be a positive number or zero.

3. Operator in hereditary operation
   a. *Selection.* Selection means to choose excellent individuals from the population as the parent to procreate offspring. With the selection standard of survival of the fittest, selection is an operation to choose the superior and eliminate the inferior by means of GA.
   b. *Crossover.* After the selection of parents for the next generation, recombination should be conducted to generate new individuals. Crossover is to exchange partial information between two individuals, and get two new ones. With the aid of selection, the species can evolve toward the expectation.
   c. *Mutation.* It refers to the process of taking the negative form of the chosen genes and generating a new individual. A very essential step is to accomplish the mutation at a quite low speed in GA. The mutation operation is preceded with bit, that is to say, to mutate a bit in the coding of an individual gene in a species. The aim of mutation is to maintain the diversity of an individual gene of a species, and to avoid a too quick ending of the search process caused by a local optimized solution.
4. Operating parameters in GA
   There are four parameters that need to be set in advance in GA:
   a. Population size. It refers to the number of individual genes in a species, and generally it is 20–100.
   b. The termination evolutionary algebra of GA. In general it is 100–500.
   c. Crossover probability. Generally it is 0.4–0.99.
   d. Mutation probability. Generally it is 0.0001–0.1.

To search for the reduct of a rough set with GA, the binary symbol string can be used to represent the population, and every binary string can code a conditional attribute subset $P \subseteq C$, which is a probable reduct, and every attribute in the subset is represented by a bit in the binary string, then we can determine whether an individual is a real reduct, if not, how long the distance is to a reduct. Choose genotype individuals through the way of random process of two extremes, and reconstruct them to generate a new genotype individual of better-expected approximation, repeat the process to thc point of satisfying the standard of ending. Because the attribute reduct of the rough set has no fixed termination condition (no ideal model of result), the concrete termination function cannot be offered. If a threshold value is set, when the difference of average fitness of several consecutive is within the value, a satisfied attribute reduct is thought to be found, and the GA can be terminated.

2.5 Application Case

In this chapter, rough set theory is applied in dealing with prediction of failure of Chinese-listed limited companies. As a worldwide topic, the prediction of company failure has long been a main research field in the theory and business circles;

thus, obviously, it is very important for investors, creditors, employees, and managers. Company failure is a universal term: one extreme definition is liquidation, while the other is the reported profit is lower than expected, and there may be various accurate definitions between the two extremes. Obviously, company failure contains bankruptcy, and it means the creditors' liquidation or receiver appointment for a single company. However, more extensively, company failure includes situations such as financial difficulties, and we can regard special treatment (ST) and particular transfer (PT) in the listed companies as company failure under the actual situation in China. ST system has been applied in China's stock market officially since 1998, and ST refers to company with abnormal finance, that is, the company suffers loss in two successive years or the net asset is less than the par value of stock. PT can be regarded as the extreme embodiment of company failure. In February 2001, China Securities Regulatory Commission issued The Implementing Procedures to Suspend and Terminate Listing of Loss-making Listed Companies, in which, the PT rule was cancelled, but if a company loses money for three successive years, its stock would be suspended.

2.5.1 Data Collecting and Variable Selection

The data comes from Shenzhen and Shanghai stock markets' annual reports of listed companies. By the end of 2001, there had been totally 1115 A-share companies, including 63 ST and PT companies. From these companies, we randomly extract 30 companies that failed (ST or PT companies), and 30 well-run companies as checking samples to form training sample group; thus we can construct decision rules. As to the rest of the companies, we randomly select 15 failures and 15 good companies to function as simulation sample groups to verify decision rules. Decision attribute d is the company's categorical attribute variable, where, the code of failure company is 1, while that of good company is 2. Totally, 8 conditional attribute variables are collected, among which a_1 is total equity (10,000 shares), a_2 is the circulating stock (10,000 shares), a_3 is the earning per share (yuan), a_4 is net asset per share (yuan), a_5 is the cash flow per share (yuan), a_6 is return on net assets (percent), a_7 is the main business income (10,000 yuan), a_8 is the net profit (10,000 yuan). The data collected is shown in Table 2.20.

2.5.2 Data Discretization

Referring to the algorithm of ID3, we can discrete automatically the data in Table 2.20 with the heuristic algorithm based on entropy. The basic idea is that dividing circularly every attribute's value set can optimize the local metric of entropy, and the criterion of the smallest descriptive length has determined the termination standard of the division process. The concrete algorithm depiction is as follows:

Table 2.20 Shenzhen and Shanghai Listed Companies' Annual Training Sample Group in 2001

Stock Name	a_1	a_2	a_3	a_4	a_5	a_6	a_7	a_8	d
Beida Hi-tech	8,398	4,166	0.132	1.141	0.062	11.58	6,198.45	1,109.51	2
ST Dasheng Electronics	14,359	7,956	0.11	1.02	0.22	10.49	33,971.51	1,533.67	1
ST PRD Co., Ltd.	54,180	9,139	0.159	0.528	0.247	30.22	102,163.94	8,638.59	1
ST Teli A	22,028	3,429	0.02	1.21	0.26	1.93	70,324.49	514.41	1
NORINCO International	10,152	2,600	0.17	3.26	–0.39	5.2	67,318.26	1,723.71	2
Ansu Cooperation	9,650	3,600	0.15	4.96	0.21	3	13,261.15	1,448.90	2
ST Xinguang	38,093	11,292	–1.15	0.13	0	–910.3	2,011.72	–43738.79	1
PT Kaidi	11,525	3,415	0.082	0.642	0.037	12.78	2,402.53	945.85	1
Gem A	38,300	4,509	0.103	1.43	–0.218	7.2	6,238.82	3,939.10	2
ST Zhangjiajie	18,360	7,436	0.095	1.14	0.83	9.55	4,946.21	1,738.90	1
ST Macro	57,568	20,371	–1.7	0.428	–0.63	–397.1	112,282.00	–97,855.67	1
PT Min Mindong	12,193	4,546	0.09	–0.15	0.43	55.21	9,256.72	1,042.94	1
TCL Communication	18,811	8,145	0.115	1.494	2.17	7.67	276,901.82	2,154.38	2
PT Jilin Light Industry	16,951	8,503	0.133	0.04	–0.05	333.81	138.17	2,258.54	1
ST Sundiro A	73,606	33,716	–0.2	1.14	0.27	–17.89	14,316.43	–150,303.47	1
PT Donghai A	36,410	4,510	0.005	0.02	0.012	24.24	2,265.16	178.22	1

(*continued*)

Table 2.20 (continued) Shenzhen and Shanghai Listed Companies' Annual Training Sample Group in 2001

Stock Name	a_1	a_2	a_3	a_4	a_5	a_6	a_7	a_8	d
Biocause	51,721	14,879	0.07	1.33	0.072	5.41	35,601.21	3,722.71	2
Baoding Swan	32,080	9,750	0.075	3.06	0.13	2.456	51,039.16	2,411.21	2
Baolihua	19,350	5,850	0.17	1.66	0.11	9.98	28,886.28	3,210.32	2
ST Hecheng	19,250	4,950	0.02	1.5	0.18	1.64	27,127.00	473.41	1
Baoshang Group	16,034	11,277	0.202	4.131	0.421	4.89	62,945.15	3,236.79	2
ST Xihuaji	6,612	1,755	0.026	0.53	0.85	4.85	5,994.25	167.14	1
Ankai Bus	22,100	10,140	0.046	2.98	0.53	1.56	44,589.60	1,024.20	2
Acheng Relay Co., Ltd.	17,555	7,892	0.23	3.13	0.2	7.27	24,709.43	3,995.21	2
AT & M	24,416	9,600	0.2768	4.333	0.117	6.39	43,509.96	6,757.37	2
ST Lianyi Microelectronics Co., Ltd.	13,686	5,158	0.0679	0.205	0.294	33.23	15,038.06	930.37	1
Cntic-trading	13,035	3,900	0.3762	4.22	2.29	9.13	248,281.10	4,903.39	2
PT Hongguang Electrical Electricity	23,000	13,573	0.035	0.006	0.212	621.15	8,605.78	813.79	1
CTV Media	23,673	7,800	0.113	3.05	0.26	3.72	36,005.81	2,682.21	2
ST Hengtai	11,545	3,500	0.106	1.098	0.521	9.46	11,465.39	1,222.52	1
Chongqing Luqiao	31,000	9,000	0.2261	2.87	0.26	7.86	13,688.80	7,007.69	2

ST Chengbai Group	7,098	2,750	0.018	0.204	0.163	9.02	64,171.88	130.4	1
Chongqing Beer	17,087	5,200	0.201	3.37	0.58	6.03	46,671.03	3,428.72	2
China Sports Industry Group Co., Ltd.	25,365	7,605	0.0561	3.063	0.29	1.83	20,690.55	1,421.87	2
ST Shenyang New District Development and Construction	19,000	7,000	0.01	1.91	−0.42	0.6	11,137.66	219.63	1
Cahic	39,000	12,000	0.23	2.13	0.58	10.78	152,495.52	8,914.16	2
Zhongsheng Pharmaceutical Co., Ltd.	13,615	3,500	0.2136	3.059	0.404	6.98	13,608.20	2,908.56	2
Chongqing Gangjiu	22,839	8,600	0.191	3.1	0.083	6.16	10,296.62	4,358.24	2
Zhongnong Resources	25,220	8,000	0.0242	2.993	−0.695	0.81	187,123.56	610.49	2
Zhongxin Pharmaceuticals	36,965	4,000	0.17	4.09	0.43	4.26	152,193.90	6,444.10	2
Sinochem International Corporation	37,265	12,000	0.24	3.87	1.73	6.28	656,715.71	9,052.08	2
China Railway Erju Co., Ltd.	41,000	11,000	0.4	4.16	0.12	9.62	403,897.71	16,415.19	2
ST Chinese Textile Machinery	35,709	2,574	0.02	0.12	0.13	15.58	44,941.93	661.2	1
ST Jiabao Group	33,369	11,520	−0.448	1.333	0.041	−33.65	19,080.59	−14,963.79	1
CDC	44,787	17,767	0.613	3.143	−0.417	19.52	76,154.61	22,564.91	2
ST Gao Sida	14,033	8,605	−0.562	0.458	0.002	−122.7	417.97	−7,881.24	1

(continued)

Table 2.20 (continued) Shenzhen and Shanghai Listed Companies' Annual Training Sample Group in 2001

Stock Name	a_1	a_2	a_3	a_4	a_5	a_6	a_7	a_8	*d*
China Enterprise	58,121	26,547	0.23	2.23	1.47	10.45	159,550.84	13,522.40	2
PT Bai Hua Cun Co., Ltd.	9,480	4,650	0.05	1.29	0.34	3.94	3,908.02	470.36	1
Chongqing Best	20,400	5,100	0.26	2.08	0.37	13.52	235,829.30	5,824.42	2
Start	35,156	13,273	0.1459	1.645	−0.094	8.8687	289,686.35	5,129.89	2
ST Huayuan	31,280	7,820	−0.077	0.327	0.0668	−23.58	28,740.87	−2,415.09	1
ST Bingxiong	12,800	3,200	0.04	0.82	−0.06	5.14	10,015.85	539.08	1
ST Ningbo Huatong	9,288	2,678	0.003	0.85	0.09	0.36	20,816.15	28.24	1
ST Zhongxi Pharmaceutical Co., Ltd.	21,559	8,423	−0.395	0.477	−0.113	−82.78	23,556.78	−8,509.91	1
PT Baoxin	26,224	1,320	0.147	1.002	0.008	14.66	37,645.00	3,884.00	1
ST Automation Instrumentation Co., Ltd.	39,929	3,367	0.02	0.28	0.25	7.59	57,844.80	835.59	1
ST BOhai	12,135	6,795	−0.277	0.885	−0.05	−31.34	2,775.29	−3,363.52	1
Jonjee Hi-tech	45,006	24,924	0.144	2.717	0.096	5.3	83,630.95	6,483.64	2
ST Jifa Express	23,490	11,532	−0.601	2.11	0.08	−28.53	87,483.63	−14,111.44	1
China Shipping Haisheng Co., Ltd.	31,728	150,609	0.15	2.5	0.23	6.08	46,762.46	4,826.88	2

Suppose heuristic entropy function of the example set $X = \{X_1, X_2, \ldots, X_m\}$ is

$$E(X) = \sum_{i=1}^{m} P_i \log m\left(\frac{1}{P_i}\right) = -\sum_{i=1}^{m} P_i \log m P_i$$

where $P_i = |X_i| \Big/ \sum_{j=1}^{m} |X_j|$, $|X_j|$ is the number of training objects and $j = 1, 2, \ldots, m$.

Obviously, informative entropy $E(X)$ relies on the number of training objects used for sample decision, and the interval of returning value is [0, 1], if the attribute set is $\{a_1, a_2, \ldots, a_n\}$, the attribute value a_i is $a_i = \{a_{i1}, a_{i2}, \ldots, a_{iv}\}$. The attribute a_i divides the training set X into $X_{i1}, X_{i2}, \ldots, X_{iv}$, where X_{ij} is the training objects set of attribute value $a_i = a_{ij}$.

When dividing a set X with attribute a_i, we can define the heuristic informative entropy function for attribute evaluation as

$$E(X, a_i) = \sum_{j=1}^{v} w_i \times E(X_{ij})$$

where $w_i = \sum_{k=1}^{m} |X_{ijk}| \Big/ \sum_{k-1}^{m} |X_k|$ and $|X_{ijk}|$ is the object number of training set X_{ij} based on the attribute value $a_i = a_{ik}$, namely, w_i is the ratio of the case number in X_{ij} to that in X.

The attribute with the least value $E(X, a_i)$ is the most appropriate attribute in dividing the intervals. The attribute value interval of numerical type is expressed as [minimum … maximum]. The interval can be divided into two parts according to the position the training objects occupy in the interval, then all the training objects are ranked in the sequence of value. The least entropy function $E(X, a_i)$ is functioned as the threshold value to evaluate all the candidate objects. After that, divide the divided segments further, until the result meets the termination criterion. When this method is applied to search for joints, the missing value is ignored. If a joint is not found for an attribute, the very attribute will not be handled.

To discrete the data in Table 2.20 automatically with the heuristic algorithm based on entropy, the discretization interval is shown in Table 2.21.

2.5.3 Attribute Reduct

The quick reduct algorithm cannot make sure that the generated attribute set is the minimal one. Considering this disadvantage, we have designed an improved quick reduct algorithm on the basis of quick reduct algorithm. The basic idea is: remove the attributes one by one from the whole conditional attribute set C, if the classification quality does not change in the division process, the procedure can

Table 2.21 Discretization Interval of Conditional Attributes

a_1—*Total Stock (10 Thousand Stock)*	a_2—*Circulating Stock (10 Thousand Stock)*	a_3—*Earning per Share (Yuan)*	a_4—*Net Asset per Share (Yuan)*
[*, 12,918)	[*, 3,550]	[*, 0.062)	[*, 1.382)
[12,918, 13,651)	[3,550, 4,510]	[0.062, 0.069)	[1.382, 1.497)
[13,651, 15,197)	[4,510, 5,025]	[0.069, 0.079)	[1.497, 1.573)
[15,197, 16,493)	[5,025, 5,129]	[0.079, 0.099)	[1.573, 1.785)
[17,019, 17,958)	[5,129, 5,179]	[0.099, 0.105)	[1.785, 1.995)
[17,958, 18,586)	[5,179, 6,323)	[0.105, 0.112)	[1.995, 2.095)
[18,586, 18,906)	[6,323, 7,521)	[0.112, 0.133)	[2.095, 2.120)
[18,906, 19,300)	[7,521, 7,810)	[0.133, 0.139)	[2.120, *)
[19,300, 20,980)	[7,810, 7,856)	[0.139, 0.147)	
[20,980, 22,064)	[7,856, 7,924)	[0.147, 0.149)	
[22,064, 22,920)	[7,924, 7,978)	[0.149, 0.155)	
[22,920, 23,582)	[7,978, 8,284)	[0.155, 0.165)	
[23,582, 25,795)	[8,284, 8,552)	[0.165, *)	
[25,795, 28,612)	[8,552, 8,603)		
[28,612, 31,140)	[8,603, 8,803)		
[31,140, 31,504)	[8,803, 9,070)		
[31,504, 32,725)	[9,070, 9,370)		
[32,725, 34,263)	[9,370, 11,285)		
[34,263, 35,433)	[11,285, 11,766)		
[35,433, 36,688)	[11,766, 13,423)		
[36,688, 37,679]	[13,423, 14,226)		
[37,679, 38,197]	[14,226, 19,069]		
[38,197, 39,465]	[19,069, 22,648]		
[39,465, 40,465]	[22,648, 30,132]		
[40,465, 52,951]	[30,132, *]		
[52,951, 57,845]			
[57,845, 65,864]			
[65,864, *]			

Table 2.21 (continued) Discretization Interval of Conditional Attributes

a_5—*Cash Flow per Share (Yuan)*	a_6—*Return on Net Assets (Percent)*	a_7—*The Main Business Income (10 Thousand Yuan)*	a_8—*The Net Profit (10 Thousand Yuan)*
[*, 0.476)	[*, 1.88)	[*, 10,717)	[*, 1166.02)
[0.476, 0.526)	[1.88, 2.20)	[10,717, 12,363)	[1166.02, 1322.20)
[0.526, 0.705)	[2.20, 3.83)	[12,363, 14,002)	[1322.20, 1491.29)
[0.705, 1.160)	[3.83, 4.10)	[14,002, 19,885)	[1491.29, 1628.69)
[1.160, *)	[4.10, 4.56)	[19,885, 20,753)	[1628.69, 1731.31)
	[4.56, 4.87)	[20,753, 24,133)	[1731.31, 1946.64)
	[4.87, 5.02)	[24,133, 25,918)	[1946.64, 2206.46)
	[5.02, 5.17)	[25,918, 28,813)	[2206.46, 2334.88)
	[5.17, 7.43)	[28,813, 31,428)	[2334.88, 3803.36)
	[7.43, 7.63)	[31,428, 34,786)	[3803.36, 3911.55)
	[7.63, 8.95)	[34,786, 36,825)	[3911.55, 7823.14)
	[8.95, 9.08)	[36,825, 40,577)	[7823.14, 8776.38)
	[9.08, 9.30)	[40,577, 44,765)	[8776.38, *)
	[9.30, 9.59)	[44,765, 45,806)	
	[9.59, 10.47)	[45,806, 54,442)	
	[10.47, 10.64)	[54,442, 60,395)	
	[10.64, 12.18)	[60,395, 63,558)	
	[12.18, 13.15)	[63,558, 65,745)	
	[13.15, 14.09)	[65,745, 68,821)	
	[14.09, 17.55)	[68,821, 73,239)	
	[17.55, 21.88)	[73,239, 85,557)	
	[21.88, *)	[85,557, 132,238)	
		[132,238, *]	

be continued until the classification quality changes when an arbitrary attribute is removed. The improved quick reduct is described as follows:

```
Improved quick reduct (C, D)
Input: The entire conditional attribute set C; the entire
  decision attributes set D.
Output: attribute reduct R ⊆ C
R←C
    Do S←R
For every conditional attribute c_i ∈ R
   If γ_{R-{c_i}}(D) = γ_S(D) then
S ← R-{c_i}
    R_1← R
      R←S
   Until γ_R(D) < γ_C(D) ∨ R = R_1
   Return R
S ← R-{c_i}
```

According to the improved quick reduct algorithm, after the reduct of the discreted training sample in Table 2.21, we get a least reduct—$RED(C)$ = {a_3 (earning per share), a_4 (net asset per share)}

2.5.4 Rule Generation

The rules set generated from the reduct $RED(P) = \{a_3, a_4\}$ is shown in Table 2.22.

2.5.5 Simulation of the Decision Rules

We can apply the simulation sample set composed of 15 failure companies and 15 non-failed companies to simulate the generated decision rules. As for those companies not accurately matching the condition of decision rules, we should calculate the distance between every rule's conditional attribute and every new object's corresponding attribute with distance measurement formula, and classify it in the nearest decision category.

If the conditional attribute value of the given object x is $a_1(x), a_2(x), \ldots, a_m(x)$, $m \leq |C|$, then the distance of object x measured by rule y is

$$D = \frac{1}{m}\left\{\sum_{l=1}^{m}\left(\frac{a_l(x) - a_l(y)}{v_{l\max} - v_{l\min}}\right)^2\right\}^{1/2}$$

where

$v_{l\max}$, $v_{l\min}$ is the maximum and minimum value of a_l, respectively
m is the number of conditional attribute

Table 2.22 Rules Set

Division Rules	*Supporting Count*
Earning per share (yuan)([0.079, 0.099]) AND net asset per share (yuan) ([*, 1.382]) ⇒ classification (1)	3
Earning per share (yuan) ([0.133, 0.139]) AND net asset per share (yuan) ([*, 1.382]) ⇒ classification (1)	1
Earning per share (yuan) ([*, 0.062]) AND net asset per share (yuan) ([*, 1.382]) ⇒ classification (1)	18
Earning per share (yuan) ([*, 0.062]) AND net asset per share (yuan) ([1.497, 1.573]) ⇒ classification (1)	1
Earning per share (yuan) ([0.062, 0.069]) AND net asset per share (yuan) ([*, 1.382]) ⇒ classification (1)	1
Earning per share (yuan) ([0.147, 0.149]) AND net asset per share (yuan) ([*, 1.382]) ⇒ classification (1)	1
Earning per share (yuan) ([*, 0.062]) AND net asset per share (yuan) ([1.785, 1.995]) ⇒ classification (1)	1
Earning per share (yuan) ([0.105, 0.112]) AND net asset per share (yuan) ([*, 1.382]) ⇒ classification (1)	2
Earning per share (yuan) ([0.155, 0.165]) AND net asset per share (yuan) ([*, 1.382]) ⇒ classification (1)	1
Earning per share (yuan) ([0.165, *]) AND net asset per share (yuan) ([2.120, *]) ⇒ classification (2)	15
Earning per share (yuan) ([0.149, 0.155]) AND net asset per share (yuan) ([2.120, *]) ⇒ classification (2)	2

(continued)

Table 2.22 (continued) Rules Set

Division Rules	*Supporting Count*
Earning per share (yuan) ([0.099, 0.105]) AND net asset per share (yuan) ([1.382, 1.497]) ⇒ classification (2)	1
Earning per share (yuan) ([0.112, 0.133]) AND net asset per share (yuan) ([1.382, 1.497]) ⇒ classification (2)	1
Earning per share (yuan) ([0.069, 0.079]) AND net asset per share (yuan) ([*, 1.382]) ⇒ classification (2)	1
Earning per share (yuan) ([0.069, 0.079]) AND net asset per share (yuan) ([2.120, *]) ⇒ classification (2)	1
Earning per share (yuan) ([0.165, *]) AND net asset per share (yuan) ([1.573, 1.785]) ⇒ classification (2)	1
Earning per share (yuan) ([*, 0.062]) AND net asset per share (yuan) ([2.120, *]) ⇒ classification (2)	3
Earning per share (yuan) ([0.112, 0.133]) AND net asset per share (yuan) ([2.120, *]) ⇒ classification (2)	1
Earning per share (yuan) ([0.165, *]) AND net asset per share (yuan) ([1.995, 2.095]) ⇒ classification (2)	1
Earning per share (yuan) ([0.139, 0.147]) AND net asset per share (yuan) ([1.573, 1.785]) ⇒ classification (2)	1
Earning per share (yuan) ([0.112, 0.133]) AND net asset per share (yuan) ([*, 1.382]) ⇒ classification (2)	1
Earning per share (yuan) ([0.139, 0.147]) AND net asset per share (yuan) ([2.120, *]) ⇒ classification (2)	1

The result shows that there are 23 objects classified correctly and 7 objects classified wrongly in 30 simulation samples; the percentage of right classification is 76.7 percent.

In the prediction of company failure, the rough set has many good properties such as the improvement of information quality, clear construction of prediction rules, and so on, which can be applied in an expert system. In addition, the generation rules of rough set is relatively simple, comprehensive, and intuitionistic. For example, the user need not possess the technical knowledge and the professional knowledge related to the classifying model explanation, and these characteristics is pretty useful for users.

2.6 Summary

This chapter mainly introduces the fundamental concepts, methods, approximation classification algorithm, and the application of rough set theory. The method of rough set theory is one of the normative methods that can convert data into knowledge. The data set is often complex and unsystematic, while knowledge is not. Knowledge can be expressed as rules of classification or decision rules. An essential principle based on rough set learning system is to find the redundancy and dependence among the given attributes in the classification problems. Rough set theory is established on the hypothesis that every object of the universe is associated with some information; meanwhile, the objects described by the same amount of information are indiscernible. The set composed of the indiscernible objects is an elementary set, and the indiscernible relation produced in this way is the mathematical foundation of RST. The main concepts of RST are approximation space, upper approximation, and lower approximation. Approximation space is a classification in which the interested domain value is divided into disjoint categories, and the categories express canonically the knowledge of the scope of domain value, while the knowledge can be understood as capacity for describing the classification characteristics of its very catalogue in the approximation space. For example, according to the characteristics of domain value of the object, the indiscernible objects belong to the same type, which means the membership relations among any arbitrary subsets in universe cannot always be defined clearly. The rough set theory uses upper approximation and lower approximation to approximate a given concept, and lower approximation is an object set whose objects absolutely belong to a subset, while the upper approximation is an object set whose objects probably belong to a subset. Any subset defined by upper approximation and lower approximation is called a rough set. The key concepts of rough set theory are "information granularity" and "reduct." Information granularity standardizes the expression of the object's limited precision; reduct represents the important properties of an information system (in accordance with objects and characteristics). The minimal independent attribute subset, which can ensure right classification and has no redundant properties, is called reduct, while the intersection of all reducts is core, which cannot be eliminated without influencing the capacity of approximation classification and is the most important property of information system. Thus, it leads to the decrease of time on compacting and calculating of data, and the acquirement of rules that the form is "if–then" from decision tables.

Chapter 3

Hybrid of Rough Set Theory and Probability

The classification done by classical rough set model must be completely correct or positive, because it is classified by strictly observing the equivalent relations and the classification is therefore accurate, namely, an object belongs to or not to a class, while there is no to-a-certain-extent "belonging to" or "contained" by a category. Classical rough set model is not able to recognize the non-decision relations, for example, to derive the predictive rules with <1 probabilities. In practical applications, the data of the knowledge base is often acquired by means of random or statistical methods, and it is likely to be noisy, ambiguous, and incomplete, so that the patterns of classes often overlap, which is not sufficient to produce deterministic rules. The probability is an objective reflection of uncertain random events. The rough set theory (RST) deals with rough concepts that do not overlap. At the same time, the roughness does not depend on event occurrence while probability depends on event occurrence. So, the hybrid of rough set model and probability may identify strong nondeterministic rules applicable for the estimation of decision probabilities from the noisy data.

3.1 Rough Membership Function

Definition 3.1 If decision table is $S = (U, A, V, f)$, and $A = C \cup D$, $C \cap D = \phi$, then the rough membership function of the decision classes is

$$\mu_X(x) = \frac{|I(x) \cap X|}{|I(x)|} \tag{3.1}$$

Rough membership function $\mu_X(x)$ can be interpreted as conditional probability, and expressed as $P(X/I(x))$, and also it can be interpreted as the confidence that x belongs to X. According to formula (3.1), rough membership function value is calculated from the data obtained, while the fuzzy set membership function is obtained by the subjective assumption.

RST with rough membership function can be approximately defined as

$$\underline{apr}(X) = \bigcup\{x \in U : \mu_X(x) = 1\} \tag{3.2}$$

$$\overline{apr}(X) = \bigcup\{x \in U : \mu_X(x) > 0\} \tag{3.3}$$

$$neg(X) = \bigcup\{x \in U : \mu_X(x) = 0\} \tag{3.4}$$

$$bnd(X) = \bigcup\{x \in U : 0 < \mu_X(x) < 1\} \tag{3.5}$$

Example 3.1 A discretization decision table is applied to calculate the rough membership functions of the decision classes, as shown in Table 3.1.

Solution: The universe is partitioned into following equivalence classes:

$$\frac{U}{C} = \{X_1, X_2, X_3, X_4, X_5\}$$

where $X_1 = \{n_1, n_3\}$, $X_2 = \{n_2\}$, $X_3 = \{n_4, n_6, n_8\}$, $X_4 = \{n_5\}$, $X_5 = \{n_7\}$;

Table 3.1 A Discretization Table of the Decision-Making

	C														
U	a_1	a_2	a_3	a_4	a_5	a_6	a_7	a_8	a_9	a_{10}	a_{11}	a_{12}	a_{13}	a_{14}	d
n_1	6	6	7	7	9	7	1	1	0	1	3	2	6	0	0
n_2	1	5	7	6	9	9	0	0	0	0	1	5	1	0	0
n_3	6	6	7	7	9	7	1	1	0	1	3	2	6	0	0
n_4	3	7	7	7	5	2	0	0	0	0	1	3	1	2	1
n_5	3	7	7	7	8	2	0	0	0	0	1	7	1	9	1
n_6	3	7	7	7	5	2	0	0	0	0	1	3	1	2	1

$$\frac{U}{D} = \{Y_1, Y_2, Y_3\}$$

where $Y_1 = \{n_1, n_2, n_3\}$, $Y_2 = \{n_4, n_5, n_6\}$, $Y_3 = \{n_7, n_8\}$.

Then, the rough membership functions of the decision classes Y_1, Y_2, Y_3 in different conditions are as follows:

$$P\left(\frac{Y_1}{X_1}\right) = \frac{|X_1 \cap Y_1|}{|X_1|} = \frac{|\{n_1, n_3\}|}{|\{n_1, n_3\}|} = 1$$

$$P\left(\frac{Y_1}{X_2}\right) = \frac{|X_2 \cap Y_1|}{|X_2|} = \frac{|\{n_2\}|}{|\{n_2\}|} = 1$$

$$P\left(\frac{Y_1}{X_3}\right) = \frac{|X_3 \cap Y_1|}{|X_3|} = \frac{|\{\phi\}|}{|\{n_4, n_6, n_8\}|} = 0$$

$$P\left(\frac{Y_1}{X_4}\right) = \frac{|X_4 \cap Y_1|}{|X_4|} = \frac{|\{\phi\}|}{|\{n_5\}|} = 0$$

$$P\left(\frac{Y_1}{X_5}\right) = \frac{|X_5 \cap Y_1|}{|X_5|} = \frac{|\{\phi\}|}{|\{n_7\}|} = 0$$

$$P\left(\frac{Y_2}{X_1}\right) = \frac{|X_1 \cap Y_2|}{|X_1|} = \frac{|\{\phi\}|}{|\{n_1, n_3\}|} = 0$$

$$P\left(\frac{Y_2}{X_2}\right) = \frac{|X_2 \cap Y_2|}{|X_2|} = \frac{|\{\phi\}|}{|\{n_2\}|} = 0$$

$$P\left(\frac{Y_2}{X_3}\right) = \frac{|X_3 \cap Y_2|}{|X_3|} = \frac{|\{n_4, n_6\}|}{|\{n_4, n_6, n_8\}|} = \frac{2}{3} = 0.667$$

$$P\left(\frac{Y_2}{X_4}\right) = \frac{|X_4 \cap Y_2|}{|X_4|} = \frac{|\{n_5\}|}{|\{n_5\}|} = 1$$

$$P\left(\frac{Y_2}{X_5}\right) = \frac{|X_5 \cap Y_2|}{|X_5|} = \frac{|\{\phi\}|}{|\{n_5\}|} = 0$$

$$P\left(\frac{Y_3}{X_1}\right)=\frac{|X_1 \cap Y_3|}{|X_1|}=\frac{|\{\phi\}|}{|\{n_1,n_3\}|}=0$$

$$P\left(\frac{Y_3}{X_2}\right)=\frac{|X_2 \cap Y_3|}{|X_2|}=\frac{|\{\phi\}|}{|\{n_2\}|}=0$$

$$P\left(\frac{Y_3}{X_3}\right)=\frac{|X_3 \cap Y_3|}{|X_3|}=\frac{|\{n_8\}|}{|\{n_4,n_6,n_8\}|}=\frac{1}{3}=0.333$$

$$P\left(\frac{Y_3}{X_4}\right)=\frac{|X_4 \cap Y_3|}{|X_4|}=\frac{|\{\phi\}|}{|\{n_5\}|}=0$$

$$P\left(\frac{Y_3}{X_5}\right)=\frac{|X_5 \cap Y_3|}{|X_5|}=\frac{|\{n_7\}|}{|\{n_7\}|}=1$$

Therefore, there is a close connection between the rough approximation and fuzzy set, and between the uncertainties and rough membership in RST. To derive strong nondeterministic rules with estimates of decision probabilities, the overlapping degree should be taken into consideration. Ziarko extended the idea of rough membership function and put forward a probability method of the rough set—variable precision rough set (VPRS).

3.2 Variable Precision Rough Set Model

The definition of rough set did not make full use of the statistic information of the border region. To solve the problem, Ziarko put forward the VPRS model by introducing the threshold value β and approximation space to reflect this kind of restrictions. VPRS is the extension of the standard RST, and it relaxes the strict definition of approximate border defined by standard precision rough set by setting the threshold parameters, meanwhile VPRS allows the classification of the probability. Compared with the standard rough set, there exists a confidence in the right classification when the objects are classified in the VPRS, which, on the one hand, perfects the conception of the approximation space and, on the other hand, helps to find out the relevant information from what is considered irrelevant according to the RST. Its main task is to solve the data classification in case of nonfunction among attributes or uncertain relationships.

When an object is classified by the VPRS, it is necessary to define a confident threshold value β. Ziarko considered β as classification errors, which are defined to be in the universe $0 \le \beta < 0.5$. However, An and other scholars defined β as the ratio of the correct classification, in which case the appropriate range is $0.5 < \beta \le 1.0$. They referred to this technique as enhanced rough set. During further research, VPRS model was extended to incorporate asymmetric bounds on the uncertain classification probability. Without loss of generality, this book is restricted to initial VPRS version and defines β as $0.5 < \beta \le 1$.

3.2.1 β-Rough Approximation

According to the basic idea of the VPRS theory, definitions are given as follows.

Definition 3.2 For a given information system $S = (U, A, V, f)$, $A = C \cup D$, $X \subseteq U$, $P \subseteq C$, $0.5 < \beta \leq 1$, β-lower approximation and β-upper approximation of X are defined, respectively, by

$$\underline{apr}_P^\beta(X) = \bigcup \left\{ x \in U \mid \frac{|I(x) \cap X|}{|I(x)|} \geq \beta \right\} \tag{3.6}$$

$$\overline{apr}_P^\beta(X) = \bigcup \left\{ x \in U \mid \frac{|I(x) \cap X|}{|I(x)|} > 1 - \beta \right\} \tag{3.7}$$

The β-lower approximation $\underline{apr}_P^\beta(X)$ of set X can also be called β-positive region, in other words, by a given confidence threshold value β, $\underline{apr}_P^\beta(X)$ is the set in which universe U can be classified into all the element sets of the set X definitely. β-Upper approximation of the set x reflects a given threshold value β, universe U can be classified into all the element set of the set X probably.

Obviously, VPRS can identify the "belonging" or "containing" to some extent, which makes it possible for it to recognize the presence of strong nondeterministic relationships, for example, to derive the predictive rules with <1 probabilities.

Definition 3.3 β-Negative region of X and β-boundary of X are defined, respectively, by

$$neg_P^\beta(X) = \bigcup \left\{ x \in U : \frac{|I(x) \cap X|}{|I(x)|} \leq 1 - \beta \right\} \tag{3.8}$$

$$bnd_P^\beta(X) = \bigcup \left\{ x \in U : 1 - \beta < \frac{|I(x) \cap X|}{|I(x)|} < \beta \right\} \tag{3.9}$$

β-Negative $neg_P^\beta(X)$ of X reflects a given confidence threshold β, and it certainly cannot be classified to all the elements of the collection set in the universe. Border zone $bnd_P^\beta(X)$ of X reflects a given confidence threshold β, and all the element sets cannot be certainly classified into either set X or set $-X$ in the universe U.

3.2.2 Classification Quality and β-Reduct

Definition 3.4 The classification quality of VPRS is defined by

$$\gamma^{\beta}(P,D)=\frac{\left|\cup\left\{x\in U:\frac{|X\cap I(x)|}{|I(x)|}\geq\beta\right\}\right|}{|U|} \tag{3.10}$$

The $\gamma^{\beta}(P, D)$ measures the proportion that the possible correct classification knowledge is in the existing knowledge for a given value β in the universe.

Definition 3.5 An approximate reduct $red^{\beta}(C, D)$ in VPRS is defined as the minimal subset of C attribute that keeps the classification quality unchanged for the given value β, and red^{β} (C, D) has the twin properties as follows:

1. $\gamma^{\beta}(C, D) = \gamma^{\beta}(red^{\beta}(C, D), D)$.
2. The subset of $red^{\beta}(C, D)$ has no the same classification quality.

Example 3.2 Suppose β = 1 and β = 0.65, calculate lower approximation, upper approximation, the classification quality, and the β-reduct according to the decision in Table 3.1 and Example 3.1.

Solution: By dividing the universe, we can obtain the following equivalence classes:

$$\frac{U}{C}=\{X_1,X_2,X_3,X_4,X_5\}$$

where $X_1 = \{n_1, n_3\}$, $X_2 = \{n_2\}$, $X_3 = \{n_4, n_6, n_8\}$, $X_4 = \{n_5\}$, $X_5 = \{n_7\}$;

$$\frac{U}{D}=\{Y_1,Y_2,Y_3\}$$

where $Y_1 = \{n_1, n_2, n_3\}$, $Y_2 = \{n_4, n_5, n_6\}$, $Y_3 = \{n_7, n_8\}$.

1. When β = 1, the upper approximation and the lower approximation of Y_1, Y_2, and Y_3 in the decision table can be expressed as follows:

$$\underline{apr}_P^{\beta}(Y_1)=\{\{n_1,n_3\},\{n_2\}\},\quad \overline{apr}_P^{\beta}(Y_1)=\{\{n_1,n_3\},\{n_2\}\}$$

$$\underline{apr}_p^{\beta}(Y_2) = \{n_5\}, \quad \overline{apr}_p^{\beta}(Y_2) = \{\{n_5\},\{n_4,n_6,n_8\}\}$$

$$\underline{apr}_p^{\beta}(Y_3) = \{n_7\}, \quad \overline{apr}_p^{\beta}(Y_3) = \{\{n_7\},\{n_4,n_6,n_8\}\}$$

$$\gamma^{\beta}(P,D) = \frac{|\{n_1,n_3,n_2,n_5,n_7\}|}{|\{n_1,n_2,n_3,n_4,n_5,n_6,n_7,n_8\}|} = \frac{5}{8} = 0.625$$

2. When $\beta = 0.65$, the upper approximation and the lower approximation of Y_1, Y_2, and Y_3 in the decision table can be expressed as the following:

$$\underline{apr}_p^{\beta}(Y_1) = \{\{n_1,n_3\},\{n_2\}\}, \quad \overline{apr}_p^{\beta}(Y_1) = \{\{n_1,n_3\},\{n_2\}\}$$

$$\underline{apr}_p^{\beta}(Y_2) = \{\{n_5\},\{n_4,n_6,n_8\}\}, \quad \overline{apr}_p^{\beta}(Y_2) = \{\{n_5\},\{n_4,n_6,n_8\}\}$$

$$\underline{apr}_p^{\beta}(Y_3) = \{n_7\}, \quad \overline{apr}_p^{\beta}(Y_3) = \{n_7\}$$

$$\gamma^{\beta}(P,D) = \frac{|\{n_1,n_3,n_2,n_5,n_7,n_4,n_6,n_8\}|}{|\{n_1,n_2,n_3,n_4,n_5,n_6,n_7,n_8\}|} = \frac{8}{8} = 1$$

According to Definition 3.5, if $\beta = 1$ and $\beta = 0.65$, we can get the reduct $\{a_1, a_{14}\}$, and the probability decision-making rules derived from the reduct $\{a_1, a_{14}\}$ are shown in Table 3.2, where the support number refers to the number of the objects in favor of this rule.

Table 3.2 The Probability Decision-Making Rules Based on the β-Reduct $\{a_1, a_{14}\}$

Rules	*Support*	*Degree of Confidence (Percent)*
$a_1 = 6 \wedge a_{14} = 0 \xrightarrow{100\text{ percent}} d = 0$	2	100
$a_1 = 1 \wedge a_{14} = 0 \xrightarrow{100\text{ percent}} d = 0$	1	100
$a_1 = 3 \wedge a_{14} = 2 \xrightarrow{67\text{ percent}} d = 1$	3	67.7
$a_1 = 3 \wedge a_{14} = 9 \xrightarrow{100\text{ percent}} d = 1$	1	100
$a_1 = 6 \wedge a_{14} = 2 \xrightarrow{100\text{ percent}} d = 2$	1	100

3.2.3 Discussion about β-Value

If $\beta = 1$, $\underline{apr}_P^\beta(X)$ and $\overline{apr}_P(X)$ coincide with the upper approximation and the lower approximation of the standard rough set model, the VPRS model comes back to the original rough set model. For the inconsistent rules of standard RST, if the inconsistency degree is weak according to the set threshold value β, this inconsistency can be viewed to be caused by a small amount of noise in the data. The rule or the main part of the rule can still be regarded as consistency rule. However, if the inconsistency degree is strong, it is viewed that no decisive information can be derived under this condition and every data object is taken as a random rule.

β-value is negatively related to the quality of classification. There are two different directions that could be taken. In one direction, when β-value increases, the quality of classification decreases. Positive region and negative region of set X will become narrower, while the boundary region of set X will become wider. A small number of objects are classified. In the other direction, when the value of β decreases, the classification precision increases, and the positive and negative regions of set X will widen while the boundary region becomes narrow, which means most objects are classified, but possibly misclassified.

Proposition 3.1 If condition class X is not given a classification with $0.5 < \beta \leq 1$, then X is also indiscernible at any level $\beta < \beta_1 \leq 1$. In contrast, if condition class X is given a classification with $0.5 < \beta \leq 1$, then X is also discernible at any level $0.5 < \beta_2 < \beta$.

If a condition class is not given a classification for every β, a condition class X is called absolutely indiscernible or absolutely rough. In other words, if and only if $bnd_P^\beta(X) \neq \Phi$, a condition class X is absolutely rough. In contrast, those only given a classification for a range of β are called relatively rough or weak discernible. For each relatively rough set X, there is a threshold value on the β-value on which set X is discernible. Associated with each conditional class is an upper bound on the β-value. If any β-value chosen is equal or below the threshold, the set X is discernable. Otherwise, there is no opportunity for majority inclusion; hence, the set X is indiscernible. The highest of these upper bounds on the β-values is defined as β max.

Proposition 3.2 For any $0.5 < \beta \leq 1$, the following holds:

1. $\overline{apr}_P^\beta(X \cup Y) \supseteq \overline{apr}_P^\beta(X) \bigcup \overline{apr}_P^\beta(Y) \bigcup \overline{apr}_P^\beta(Y)$
2. $\underline{apr}_P^\beta(X \cap Y) \subseteq \underline{apr}_P^\beta(X) \bigcap \underline{apr}_P^\beta(Y)$
3. $\underline{apr}_P^\beta(X \cup Y) \supseteq \underline{apr}_P^\beta(X) \bigcup \underline{apr}_P^\beta(Y)$
4. $\overline{apr}_P^\beta(X \cap Y) \subseteq \overline{apr}_P^\beta(X) \bigcap \overline{apr}_P^\beta(Y)$

Proof

1. For any two $X \subseteq U$, $Y \subseteq U$, given β-value

$$\frac{|I(x) \cap (X \cup Y)|}{|I(x)|} \geq \frac{|I(x) \cap X|}{|I(x)|}$$

and

$$\frac{|I(x) \cap (X \cup Y)|}{|I(x)|} \geq \frac{|I(x) \cap Y|}{|I(x)|}$$

therefore, $\overline{apr}_P^{\beta}(X \cup Y) \supseteq \overline{apr}_P^{\beta}(X) \bigcup \overline{apr}_P^{\beta}(Y)$.

2. For any two X, $Y \subseteq U$, given β-value

$$\frac{|I(x) \cap (X \cap Y)|}{|I(x)|} \leq \frac{|I(x) \cap X|}{|I(x)|}$$

and

$$\frac{|I(x) \cap (X \cap Y)|}{|I(x)|} \leq \frac{|I(x) \cap Y|}{|I(x)|}$$

therefore, $\underline{apr}_P^{\beta}(X \cap Y) \subseteq \underline{apr}_P^{\beta}(X) \bigcap \underline{apr}_P^{\beta}(Y)$.

(3) and (4) can be proved in a similar way.

A simple illustrative example is shown below.

Proposition 3.3 When β = 1, VPRS model comes back to be classical rough set.

Proof: When β = 1, $|I(x) \cap Y|/|I(x)| \geq \beta = 1$, we can consider $I(x) \subseteq Y$, lower approximation of VPRS will be simplified as lower approximation of classical rough set; $|I(x) \cap Y|/|I(x)| > 1 - \beta = 0$, we can think $I(x) \cap Y \neq \phi$, lower approximation of VPRS will be simplified as lower approximation of classical rough set.

Table 3.3 An Information Expression System

U	Condition Attributes (C)						Decision Attribute (D) d
	a_1	a_2	a_3	a_4	a_5	a_6	
n_1	1	2	1	1	2	1	*N*
n_2	1	2	2	2	2	2	*N*
n_3	2	2	1	2	2	2	*N*
n_4	1	2	1	1	2	1	*P*
n_5	2	2	2	2	1	1	*P*
n_6	1	2	1	1	2	1	*P*
n_7	2	2	2	2	1	2	*P*

Example 3.3 An information expression system given in Table 3.3, $S = (U, A, V, f)$, where $U = \{n_1, n_2, n_3, n_4, n_5, n_6, n_7\}$, the set of condition attributes $C = \{a_1, a_2, a_3, a_4, a_5, a_6\}$, and the set of decision attributes $D = \{d\}$. Ask for β-reduct and classification quality of different threshold β max.

Solution: $U/C = \{X_1, X_2, X_3, X_4, X_5\}$
where $X_1 = \{n_1, n_4, n_6\}$, $X_2 = \{n_2\}$, $X_3 = \{n_3\}$, $X_4 = \{n_5\}$, $X_5 = \{n_7\}$;

$$\frac{U}{D} = \{Y_N, Y_P\}$$

where $Y_N = \{n_1, n_2, n_3\}$, $Y_P = \{n_4, n_5, n_6, n_7\}$.

β-Reduct and classification quality computed are shown in Table 3.4. β-Reduct in Table 3.4 can provide the probability of decision rules. For example, Table 3.5 shows the probability decision rules constructed with β-reduct $\{a_1, a_3\}$.

Table 3.4 β-Reduct and Classification Quality

β-Reduct	Classification Quality	β_{max}
$\{a_2\}$	1	0.57
$\{a_1, a_3\}$	1	0.67
$\{a_1, a_4\}$	1	0.67
$\{a_4, a_5\}$	0.57	1

Table 3.5 Decision Rules for β-Reduct {a_1, a_3}

Rules	*Support*	*Degree of Confidence (Percent)*
$a_1 = 1 \wedge a_3 = 2 \xrightarrow{\text{100 percent}} d = N$	1	100
$a_1 = 2 \wedge a_3 = 1 \xrightarrow{\text{100 percent}} d = N$	1	100
$a_1 = 1 \wedge a_3 = 1 \xrightarrow{\text{67 percent}} d = P$	3	67
$a_1 = 2 \wedge a_3 = 2 \xrightarrow{\text{100 percent}} d = P$	2	100

3.3 Construction of Hierarchical Knowledge Granularity Based on VPRS

As a convenient and effective tool, the RST is used to deal with knowledge granules. In RST, the knowledge granularity is a typical division of the universe, and some information is lost with formation of knowledge granularity; meanwhile, some subsets implied in the universe can only be approximately described and expressed by overlapping of the granularity and sets. However, the structure of knowledge granularity is single hierarchical granule structure in the RST, and too much and too small knowledge granules are formed due to the diversity of attributes, which are so complex that we cannot identify the overall characteristics of the data. In practical application, according to requirements of different problems, it is required to show the different levels knowledge granularity, that is to say, the knowledge granularity is often divided into different levels. When fewer details are asked for, we often put the similar knowledge granulation together; when more details are demanded, we need to show more detailed knowledge granulation, and then granules should be in another class, which forms the information pyramid. To deal with this kind of knowledge granular structure, we apply the VPRS model to build a hierarchical knowledge granularity in this section.

3.3.1 Knowledge Granularity

Knowledge granularity, also known as the information granulation, is a set of objects that are indiscernibly put together because of their similarity and functional proximity. Knowledge granulation is an important issue in computational science, and there are a lot of reasons for people to study it, while the problem simplification and actual needs are the major driving factors. When the issue involves incomplete, uncertain, or vague information, it may be difficult or unnecessary to discern different elements, and the knowledge granularity is used to acquire an effective tool for modularization. Sometimes, detailed information may be obtained, but we need to use the knowledge granularity to get efficient and practical solutions. It may not be necessary to use very precise solution in many cases, while the application of knowledge granulation could make practical problems simplified. In many cases, it may be very expensive to acquire accurate information, while it could reduce costs by using the knowledge granularity to acquire information. The knowledge granularity plays an important role in design and implementation of intelligent information systems. No matter what basic technology is needed, the driving force behind the knowledge of granularity includes some essential elements:

1. The problem must be broken down into a series of easier and smaller subtasks. For example, when we need to comprehensively understand and obtain an effective insight into the essence of the problem, instead of including all unnecessary details, we can use the formation of knowledge granularity in abstract process of simplifying the whole concept. In fact, by changing knowledge granularity, we can hide or reveal a certain number of details that are planned to be dealt with in the process of designing.
2. The construction, expression, interpretation, and application of granularity are basic contents of the knowledge granularity.

3.3.2 Relationship between VPRS and Knowledge Granularity

The knowledge granularity is indiscernibility of the objects resulted from the lack of enough information. Therefore, there is a close relationship between knowledge granularity and indiscernibility.

3.3.2.1 Approximation and Knowledge Granularity

From the viewpoint of VPRS, the β-lower approximation of $X \subseteq U$ consists of the union of knowledge granularity that surely belongs to X, of which confidence level is less than β in the U, and the β-upper approximation of X consists of the union of knowledge granularity that surely belongs to X, of which confidence level is greater than $\beta - 1$. We can reveal the granular structure of the complex concept with approximation.

3.3.2.2 Classification Quality and Granularity Knowledge Granularity

Apparently, $0 \le \gamma^{\beta}(P, D) \le 1$, if $\gamma^{\beta}(P, D) = 1$, $\beta = 1$, that is to say, granularity of knowledge in D depends entirely on C, then it is recorded as $C \Rightarrow D$; if $\beta = 1$, $\gamma^{\beta}(P, D) < 1$, that means, part of the function of the granularity of knowledge in D depends on the granularity of knowledge of C; if $0.5 < \beta \le 1$, $\gamma^{\beta}(P, D) < 1$, that is, D approximation depends on C, relies less on the nature of approximation than the nature of functional dependency, for example, no longer existence of transitivity. $\gamma^{\beta}(P, D)$ measures the β-value of a given universe and the correct knowledge classification may account for the percentage of existing knowledge. And the interdependence between C and D, to a certain degree, reveals the degree that the knowledge granular structure of D is used by related knowledge of granularity in C.

By the formula (3.10), the value of β and the classification accuracy takes on negative correlation, and it is difficult to coordinate the balance for decision-makers. That is to say, there may be a small number of unclassified objects, while the majority of the targets may not have the correct formation of knowledge granularity; or the majority of objects are not classified, while a small number of objects may not have formed the correct knowledge granularity.

3.3.3 Construction of Hierarchical Knowledge Granularity

3.3.3.1 Methods of Construction of Hierarchical Knowledge Granularity

The ordinal relationship $I_1 \subset I_2$ can be defined with the inclusion relation of sets in the universe U; in other words, equivalence relation I_1 is more precise than I_2. According to the views of equivalent granularity, the knowledge granularity generated through more precise relationship is smaller than the knowledge granularity generated by more rough relationship, and in fact, each equivalent granulation in I_2 is the union of those in I_1, that is to say, each granularity in I_1 is obtained by dividing the granule of I_2. The above formulas can be expanded into m nested relations:

$$I_1 \subset I_2 \subset \cdots I_m$$

Corresponding equivalent knowledge granularity should meet the following condition:

$$I_1(x) \subseteq I_2(x) \subseteq \cdots I_m(x)$$

According to nested equivalence relation, the tree structure chart of knowledge granularity can be formed with different confidence threshold β and different classification quality of $\gamma^{\beta}(P, D)$, in which each node represents a class, and the root is the largest cluster composed of all the elements in U, which is divided into a group of knowledge granularities, that is to say, the formation of subclass roots is a division of root. Similarly, each cluster can be further divided into smaller disjoint clustering, that is, much fine knowledge granularity, and the leaves are equivalent to universe elements. Hierarchy can be regarded as granularity knowledge, that is, the combination of the finer knowledge granularity from the bottom to the top. On one level, all the elements in the lower-level knowledge granulations are included in a series of nodes of the nested knowledge granulation and root. Generally speaking, given an attribute set, we can construct a hierarchical knowledge granulation table, and every knowledge granulation has a corresponding subset of attributes, a higher level with a rougher granularity. In such a hierarchical structure of knowledge granularity, decision-makers can determine the appropriate knowledge granularity and the probability of the corresponding decision rules.

3.3.3.2 Algorithm Description

To construct a hierarchical knowledge granularity based on different confidence threshold β and different levels classification quality γ with nested equivalent relations set, we can use the algorithm described as follows:

Step 1: Input decision knowledge representation system $S = (U, A, V, f)$ of the discretization from raw data.

Step 2: Let the number of levels $k = 1$, $\beta_k = |I(x) \cap X| / |I(x)|$.

Step 3: Calculate each β_k of the attribute value for $a_i = v_j$ in the decision-making knowledge system.

Step 4: If the knowledge granularity with the minimum $\beta_k > 0.5$ in the attribute value for $a_i = v_j$ is not an empty set, then the knowledge granularity is classified by this attribute as the first level of knowledge granularity of the root. Calculate the classification quality γ_k that is greater than the minimum β_k, and add an attribute into Step 7, otherwise add in an attribute into the Step 5.

Step 5: Calculate each β_k of the model Λ_i $[a_i = v_j]$ in the relationship table.

Step 6: If the model Λ_i $[a_i = v_j]$ of the minimum $\beta_k > 0.5$ knowledge granularity is nonempty, then use classification of granularity of knowledge according to attributes from the model size as the first level of the root of knowledge granularity, and calculate the classification quality γ_k, which is greater than the minimum quality β_k, to increase an attribute into Step 7, otherwise an additional attribute into Step 6.

Step 7: If the knowledge granularity of the smallest $\beta_{k+1} > 0.5$ is nonempty in the model Λ_i $[a_i = v_j]$, and the minimum $\beta_{k+1} \neq$ the minimum β_k, or $\gamma_{k+1} \neq \gamma_k$, then the knowledge granularity classified by the pattern is regarded as sublayer of the current knowledge granularity, and let $k = k + 1$, repeat Step 7 by adding an attribute.

Step 8: Until the model Λ_i $[a_i = v_j]$ covers all attributes or classification quality, $\gamma = 1$.

Step 9: Construct a hierarchical knowledge granulation table on $\beta_{min} < \beta \leq \beta_{max}$ and γ.

Step 10: Construct the figure of hierarchical granularity of knowledge.

Table 3.6 A Knowledge System of Patient with Congenital Anomaly

U	Condition Attributes (C)						Decision Attribute (D) d
	a_1 (Round)	a_2 (Telorism)	a_3 (Cornea)	a_4 (Slanting)	a_5 (IrisDefects)	a_6 (Eyelashes)	
n_1	No	Normal	Megalo	Yes	Yes	Long	Aarskog
n_2	Yes	Hyper	Megalo	Yes	Yes	Long	Aarskog
n_3	Yes	Hyper	Normal	No	No	Normal	Down
n_4	Yes	Hyper	Normal	No	No	Normal	Down
n_5	Yes	Hyper	Large	Yes	Yes	Long	Aarskog
n_6	No	Hyper	Megalo	Yes	No	Long	Cat-cry

Example 3.4 A knowledge system $S = (U, A, V, f)$ of patient with symptom of congenital anomaly shown in Table 3.6, where the universe is $U = \{n_1, n_2, n_3, n_4, n_5, n_6\}$, the set of condition attributes $C = \{a_1, a_2, a_3, a_4, a_5, a_6\}$, and the set of decision attributes $D = \{d\}$. Please calculate different confidence threshold β for the hierarchical knowledge granularity.

Solution: The universe is partitioned into the following equivalence classes:

$$\frac{U}{C} = \{X_1, X_2, X_3, X_4, X_5\}$$

where $X_1 = \{n_1\}$, $X_2 = \{n_2\}$, $X_3 = \{n_3, n_4\}$, $X_4 = \{n_5\}$, $X_5 = \{n_6\}$;

$$\frac{U}{D} = \{Y_A, Y_D, Y_C\}$$

where $Y_A = \{n_1, n_2, n_5\}$, $Y_D = \{n_3, n_4\}$, $Y_C = \{n_6\}$.

According to the algorithm described in Section 3.3.3, we can derive the nested series of attribute subset $\{a_1, a_2, a_3\}$, $\{a_2, a_3\}$, $\{a_3\}$. Obviously, $I_{\{n_1,n_2,n_3\}}(p) \subseteq I_{\{n_2,n_3\}}(p) = I_{\{n_3\}}(p)$, thus we get the hierarchical knowledge granularity of different confidence threshold and different levels of classification quality, as shown in Table 3.7.

As described in Table 3.7, we can construct a hierarchical knowledge granularity table based on the different confidence threshold β and different levels of classification quality γ, as shown in Figure 3.1.

Hierarchical knowledge granularity realizes data compression and data blocks, and provides the background of the further detailed analysis of data, so it is an effective way to do date mining. The structure of hierarchical knowledge granularity based on VPRS can give appropriate knowledge granularity according to

Table 3.7 Hierarchical Knowledge Granularity Table

Set of Attributes of Classification	β	*Knowledge Granularity*	*β-Lower Approximation*	$\gamma^{\beta}(P, D)$
$\{a_3\}$	$0.67 < \beta \le 1$	$\{n_1, n_2, n_6\}$, $\{n_3, n_4\}$, $\{n_5\}$	$\{n_3, n_4\}$, $\{n_5\}$	3/6
$\{a_2, a_3\}$	$0.5 < \beta \le 1$	$\{n_1\}$, $\{n_2, n_6\}$, $\{n_3, n_4\}$, $\{n_5\}$	$\{n_1\}$, $\{n_3, n_4\}$, $\{n_5\}$	4/6
$\{a_1, a_2, a_3\}$	$\beta = 1$	$\{n_1\}$, $\{n_2\}$, $\{n_6\}$, $\{n_3, n_4\}$, $\{n_5\}$	$\{n_1\}$, $\{n_2\}$, $\{n_3, n_4\}$, $\{n_5\}$, $\{n_6\}$	1

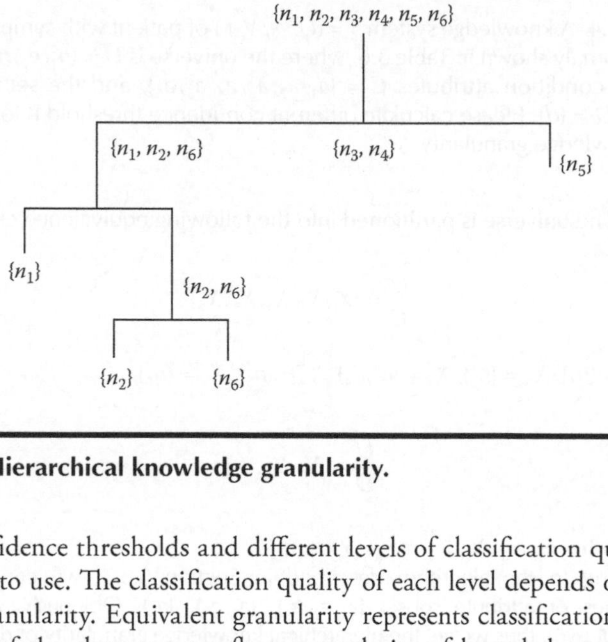

Figure 3.1 Hierarchical knowledge granularity.

different confidence thresholds and different levels of classification quality, so it is more flexible to use. The classification quality of each level depends on the size of equivalent granularity. Equivalent granularity represents classification knowledge, and the size of the element set reflects granularity of knowledge representation, while finer knowledge granularity tends to find more rules and has a higher probability of decision prediction, but the support strength of decision rules is rather weak, and these rules may not be correct; more coarse knowledge granularity may typically identify relatively weak and simple rules, and these rules generally have lower decision probability, but they are likely to be correct. As far as the concepts are concerned, some machine learning and data mining methods can be considered the search of suitable approximation of a certain structure based on granularity knowledge.

3.4 Methods of Rule Acquisition Based on the Inconsistent Information System in Rough Set

VPRS model has a error-tolerance by introducing a confidence level β. However, the model cannot identify random rules supported by only a small number of examples. This section offers a method that can be used to learn both decision-making and non-decision-making. By converting attribute-oriented method into the method that allows learning classification rules of non-decision-making, we can get more general and reliable classification rules.

3.4.1 Bayes' Probability

Bayes' probability formula is

$$P\left(\frac{B_i}{B}\right) = \frac{P(B_i)P(B/B_i)}{\sum_{i=1}^{n} P(B_i)P(B/B_i)} \tag{3.11}$$

where

$\{B_1, B_2, \ldots, B_n\}$ is a division of universe U
$P(B_i)$ is the probability of the event B_i
$P(B)$ is the sum of all possible B_i in the joint probability $P(B, B_i)$

The formula (3.11) can be described by words as

$$\text{Posterior probability} = \frac{\text{Likelihood function} \times \text{prior probability}}{\text{Evidence factor}}$$

Posterior probability equals likelihood function times priori probability and divides evidence factor.

Total probability formula is a calculation formula by which we can get results from the reasons, while Bayes' formula is used to find its reason when the result is given, therefore, Bayes' formula is also known as posterior probability formula. Bayes' formula shows that by observing the value of B, we can convert the priori probability $P(B_i)$ into posterior probability $P(B/B_i)$. That is to say, when we assume eigenvalue B is known, its category belongs to probability B_i, and $P(B/B_i)$ can be regarded as B_i on the likelihood function of B, that is to say, when all other conditions equal, the larger B_i in $P(B/B_i)$ is more likely to be true category. Posterior probability mainly is determined by the product of the a priori probability and the likelihood function, and the evidence factor $P(B)$ can be only regarded as a scalar factor, so that it can make sure that the sum of all categories of posterior probability is 1, which satisfies the probability condition.

The advantage of Bayes' theory is that it can make sure the quantitative compatibility as a result of the fact that Bayes' theory is built on the basis of axioms, while other classification methods cannot. However, there are some shortcomings; for example, if we make just a decision, then it will cause a lot of problems; there is no reasonable way to determine the prior probability. The most serious disadvantage is that it is more difficult to calculate the condition density function, while multi-Gaussian model can provide a sufficient approximation to many true densities. However, it is quite different from Gaussian model in the density form of many problems. It is not a simple matter to estimate the unknown parameters directly from the sample data, even though the Gaussian model meets the requirements. Some shortcomings of Bayes' methods can be improved by rough set method.

3.4.2 Consistent Degree, Coverage, and Support

To find the relevant rules, some scholars have put forward the concept of consistent degree, the consistent degree of decision rules *r*, also known as uncertainty factor, is represented by $cer_I(Y)$, defined as

$$cer_I(Y) = \frac{|I(x) \cap Y|}{|I(x)|} \tag{3.12}$$

where Y is the decision class $Y \subseteq U$.

The consistent degree measures the precision of distribution of the condition types to the decision types, which can be interpreted as a confidence level, similar to the proportion of correct classification β in VPRS, which has been widely used in data mining. Obviously, if $cer_I(Y) = 1$, then $CON_C(X_i) \rightarrow DEC_D(Y_j)$ is true, then we call the information systems S as decision-making or consistent, that is to say, the condition attribute only describes the decision attribute; if $0 < cer_I(Y) < 1$, then we regard S as non-decision-making or inconsistent, in other words, the condition attribute probably determines the decision attribute, which is caused by the inconsistency of information. Therefore, we can get that the uncertainty rules is uniquely supported by lower approximation of the corresponding decision-making class, while probability rules is only supported by the boundary object of the corresponding decision class.

The coverage of decision rules is defined as

$$\text{cov}_I(Y) = \frac{|I(x) \cap Y|}{|Y|} \tag{3.13}$$

$\text{cov}_I(Y)$ is used to estimate the quality of decision-making rules. Coverage reveals probabilities that satisfy the type of conditions, which can be considered the confidence level against the rules, under the decision-making conditions that meet the decision-making rules.

Consistency is the probability of decision types Y on condition of condition type $I(x)$, while coverage is the probability of condition types $I(x)$ on condition of decision type Y. Pawlak has studied Bayes' relations between consistent degree and coverage.

Suppose $cer_I(Y) = |I(x) \cap Y|/|I(x)| = P(Y/I(x))$, $cer_I(Y) = |I(x) \cap Y|/|I(x)| = P(Y/I(x))$, then $cer_I(Y) \times P(I(x)) = P(I(x)) \times P(Y/I(x)) = P(I(x),Y) = P(I(x)/Y) \times P(Y) = \mathrm{cov}_I(Y)$ $P(Y)$ $cer_I(Y)$ measures the full degree of $CON_C(X_i) \rightarrow DEC_D(Y_j)$, $\mathrm{cov}_I(Y)$ measures the necessity of $CON_C(X_i) \rightarrow DEC_D(Y_j)$. In other words, if $\mathrm{cov}_I(Y) = 1$, then $CON_C(X_i) \rightarrow DEC_D(Y_j)$ is true; if $\mathrm{cov}_I(Y) = 1$, then $DEC_D(Y_j) \rightarrow CON_C(X_i)$ is true; and if $\mathrm{cov}_I(Y) = 1$ and $\mathrm{cov}_I(Y) = 1$, then $conc_C(X_i) \leftrightarrow dec_D(Y_j)$. However, the relation between consistency and coverage need not use posterior probability, which is the basis of Bayes' analysis.

Support $\sup_I(Y)$ can be defined as

$$\sup_I(Y) = \frac{stre_I(Y)}{|U|} \tag{3.14}$$

where $stre_I(Y)$ is the support number of the rules $conc_C(X_i) \rightarrow dec_D(Y_j)$ in S, or the target number of attribute values and matching rules on the universe. Support $stre_I(Y)$ defines the strength of the decision rules.

3.4.3 Probability Rules

When it is difficult to obtain certain decision rules in practical application, the derivation of the probability rules is an important means of solving the problems. Let the consistent degree, coverage, and support thresholds to be β, α, γ, respectively, the probability rules can be defined as

$$r : conc_C(X_i) \xrightarrow{\sup_I(Y),\, cer_I(Y)} dec_D(Y_j) \tag{3.15}$$

or

$$r : conc_C(X_i) \xrightarrow{\sup_I(Y),\, \mathrm{cov}_I(Y)} dec_D(Y_j) \tag{3.16}$$

where $X_i \cap Y_j = \phi$, $\sup_I(Y) \geq \gamma$, $cer_I(Y) \geq \beta$, and $\mathrm{cov}_I(Y) \geq \alpha$. This rule is a probability rule with three statistical measures, in other words, it only searches for the probability rules in which the consistency or coverage is more than or equal to the threshold value for the categories with enough support in information systems. When the support is not taken into consideration, formula (3.15) will come back to be the rules defined by the VPRS method. When the support is not taken into consideration and $\beta = 1$, formula (3.15) will come back to be the rules defined by the rough set method.

3.4.4 Approach to Obtain Probabilistic Rules

We hope that the data in information representation system can be sufficient consistent, or covering enough condition classes or decision classes. This is a reasonable assumption, which is especially useful in a large-scale database containing many types. In the database, as for the category with few support, if they are less than a given support threshold, they will be regarded as noise data, thus it will help to save computing time and export more simple and applicable reduct. We have designed an algorithm for mining the probability rule, and the main idea is this: For each rule, first, examine whether the support is enough; if yes, then examine whether they meet the consistent degree and coverage of support type according to a pre-set threshold, and put the rules that meet the requirements into the rule base as knowledge. The detailed algorithm is described as follows:

Step 1: Calculate each pair of element value $[f(x,q_j)=r_{q_j}]$ and form a candidate set of elements L_1.

Step 2: Let $k = 1$.

Step 3: If L_k is nonempty, calculate the support of each candidate model $\wedge_k[f(x,q_j) = r_{qj}]$, meanwhile put the model with a support greater than or equal to the threshold γ into a collection of R_k.

Step 4: As for each model $\wedge_k[f(x,q_j) = r_{qj}]$ in the set R_k, calculate its consistent degree and coverage. If the consistent degree is greater than or equal to the threshold β or the coverage is greater than or equal to threshold α, then add the model into the rule base and remove it from R_k.

Step 5: If R_k is nonempty, then intersect model $\wedge_k[f(x,q_j) = r_{qj}]$ in the set R_k to get L_{k+1}.

Step 6: If the model $\wedge_k[f(x_i,q) = r_{qj}]$ is not in the set R_k, then carry out the operation of the deletion in L_{k+1}.

Table 3.8 A Knowledge Representation System of Global Warming

U	*Condition Attributes (C)*			*Decision Attribute (D) d*	*Support Number (stre$_l$(D)) cn*
	a_1	a_2	a_3		
n_1	High	Low	Low	Low	2
n_2	Medium	High	Low	High	20
n_3	High	High	High	High	30
n_4	Medium	Low	High	High	90
n_5	Low	Low	Low	Low	120
n_6	High	High	Medium	High	70
n_7	Medium	Low	High	Low	34

Step 7: Let $k = k + 1$.

Step 8: Repeat the operation from Step 3 to Step 6 until the R_k becomes an empty set.

Example 3.5 Suppose that a knowledge representation system of global warming is shown in Table 3.8, and take consistent degree, coverage, and support as β = 70 percent, α = 5 percent, and γ = 1 percent, respectively, figure out reduct and its generated rule set of the decision probability.

In Table 3.8, the meaning of attribute is follows: a_1, solar energy; a_2, volcanic activity; a_3, residual CO_2; d, temperature; cn, days count. Let β = 70 percent, α = 5 percent, γ = 1 percent, according to the algorithm described in Section 3.4.4, we get a reduct $\{a_1, a_3\}$, and a rule set of the decision probability generated from the method, as shown in Table 3.9.

Given confidence β = 70 percent, and based on VPRS model, we can get a rule set of the probability, as shown in Table 3.10.

Table 3.9 The Set of the Probability Rule

Rules	*Support (Percent)*	*Consistent Degree (Percent)*	*Coverage (Percent)*
$a_1 = \text{Medium} \wedge a_3 = \text{Low} \rightarrow d = \text{High}$	5.5	100	9.5
$a_1 = \text{High} \rightarrow d = \text{High}$	27.3	98.0	27.3
$a_1 = \text{Medium} \wedge a_3 = \text{High} \rightarrow d = \text{High}$	24.6	72.6	42.9
$a_1 = \text{Low} \rightarrow d = \text{Low}$	32.8	100	76.9

Table 3.10 A Rule Set of the Probability Generated from $\{a_1, a_3\}$ of β-Reduct

Rules	*Confidence (Percent)*
$a_1 = \text{High and } a_3 = \text{Low} \xrightarrow{\text{100 percent}} d = \text{Low}$	100
$a_1 = \text{Medium and } a_3 = \text{Low} \xrightarrow{\text{100 percent}} d = \text{High}$	100
$a_1 = \text{High and } a_3 = \text{High} \xrightarrow{\text{100 percent}} d = \text{High}$	100
$a_1 = \text{High and } a_3 = \text{Medium} \xrightarrow{\text{100 percent}} d = \text{High}$	100
$a_1 = \text{Medium} \wedge a_3 = \text{High} \xrightarrow{\text{72.6 percent}} d = \text{High}$	72.6
$a_1 = \text{Low} \xrightarrow{\text{100 percent}} d = \text{Low}$	100

By comparing Table 3.9 with Table 3.10, we can know that the rule sets of the Table 3.9 are much simpler. In Table 3.10, because the first rule is supported by only two examples and the support is 0.5 percent, so the record is not reliable, thus we regard it as noise data according to a given threshold. For example, it may happen that the "Low" attribute values of a_1 are mistakenly written as the "High" in situation of data inputting, so we regard the rule a_1 = High and a_3 = Low $\xrightarrow{\text{100 percent}}$ d = Low as a random rule, while such a rule still can be regarded as a decision rule according to VPRS methods.

3.5 Summary

This chapter first introduces rough membership function, VPRS, and some basic concepts about Bayesian probability, and then discusses further some properties about VPRS model and confidence threshold parameters β. On this basis, we discuss some problems about knowledge granularity in VPRS model and analyze knowledge granularities, from fine ones to coarse ones, at different levels. Meanwhile, we make use of the nested equivalence relations set to construct hierarchical knowledge granularity that is suitable for different confidence threshold β and classification quality γ at different levels, and design the corresponding algorithm. Considering some shortcomings of VPRS in solving practical problems, for example, the VPRS model cannot distinguish the random rules supported only by a small number of cases, we discuss a new model of hybrid of RST and probability to acquire more general and reliable classification rules. This model can not only study the decision classification rules but also the non decision-making classification rules, while the inconsistent level supported by the rules is designated by the users, and we also design the corresponding algorithm. The basic idea of the algorithm is to allow the users to designate three parameters in the process of learning—the minimal support, the consistent degree that the classification rules must satisfy, and coverage—and based on these conditions, we deduce the rules that satisfy the requirements of the parameter. This method makes full use of the inherent statistics in the knowledge system, and its features are as follows: (1) the stronger function in dealing with the noise, (2) the more reliable and simpler classification rules derived, and (3) the more effective analysis of the large-scale database.

Chapter 4

Hybrid of Rough Set and Dominance Relation

The classical rough set theory (RST) cannot be used to discover the relevant inconsistencies of the designated preference attributes in preferential multiple attribute decision table, for example, the attributes with preference information that we often encounter in economy and finance decision situation, such as return on investment, profit rate, market share, and debt ratio. Solving these problems is crucial for the application of rough set approach in multicriteria decision analysis (MCDA).

4.1 Dominance-Based Rough Set

To deal with preference attributes, Greco and other scholars have put forward the dominance-based RST, which can deal with the inconsistencies of the typical exemplary decisions in MCDA. The criterion represents the attribute with preference information. This improvement is mainly based on dominance relation instead of indiscernibility relation in the rough approximation of decision classes; thus, we can obtain a very important result that the preference model can be generated from the exemplary decisions according to the decision rules expressed by the logical statements "if…then…."

4.1.1 The Classification of the Decision Tables with Preference Attribute

The decision table includes the information that are described by some attributes and related to object set, while the analysis of such a table consists in approximating the classifications induced by decision attributes by means of the classification induced by condition attributes in the classical rough set. The establishment of these two kinds of classifications is independent, that is to say, one is not deduced from the other. The aim of the decision analysis is to answer two questions: the first one is to explain decisions according to the conditions in which they are made, while the second one is to give a suggestion about making a decision under specific circumstances, which is mainly based on decision rules induced from a decision table. In this sense, the rough set approach is similar to the inductive learning approach. The former approach, however, goes far beyond the latter because the decision suggestions are made according to the interpretation of the useful information that supports decision in the rough set approach, such as reducts, core, and approximation quality. The decision analysis can distinguish three common kinds of tasks: classification, selection, and order sorting. Generally speaking, decisions are based on some characteristics of objects, and the decisions depend on the interpretation of attributes given by decision-makers. For example, when you buy a car, your decision may be based on the characteristics such as price, maximum speed, fuel consumption, color, and producing area, and these characteristics are called attributes. Some attributes may contain preference information and the others may not. For example, price, maximum speed, and fuel consumption are preference attributes. Because, in most cases, the lower the price and the consumption of fuel are, the better; the higher the speed is, the better. Generally speaking, color and producing area may not possess preference information. Preference order may exist in decision. For example, for a catalog, the classification of automobiles does not impose any preference order among the classes (trucks, buses, taxi, etc.); however, the selection of the best automobile or the order sorting of cars from the best to the worst certainly impose a preference order. The class may express one kind of preference when it depends on interpretation given by the decision-maker for the classification. For example, the decision-maker may classify automobiles into three categories: the acceptable one, the hardly acceptable one, and the nonacceptable one. Therefore, the classification may also be ordinal—this type of classification is called order sorting.

The classical rough set approach used in decision analysis is suitable for multiattribute classification that assigns object set described by attribute set to one of the predefined categories, because the set of classification examples may be represented directly in the decision table. What's more, it is possible to extract all the essential knowledge contained in the table based on indiscernibility or similarity relations. But, the classical rough set approach cannot extract all the essential knowledge contained in the decision table of multicriteria sorting problems, for example, classification assigns to object set described by criteria set to one of the predefined and preference-ordered categories. However, in many real problems, it is very important to consider the ordinal property of preference attribute. There are two major models used in MCDA: functional and relational ones. The relational models are mostly expressed by dominating relation and fuzzy relation. These models require specific preferential information related to their parameters. People often make decisions by searching for suitable rules that provide good justification for their choices. So, after getting preference information in terms of exemplary comparisons, people will establish preference model according to "if…then…" rules. Then, these rules can be applied in a set of objects to obtain special preference relations. Based on the application of these relations, a suitable suggestion can be given to support the decision in current decision-making. Inducing rules from examples is a typical approach of artificial intelligence, and then these rules not only explain the preference behavior of the decision-maker but also provide decision suggestions for him or her.

Some attributes considered in rough set, such as symptom, color, and texture, are different from preference attributes, because their attribute values do not take preference information into consideration, but preference attributes do. The presence of preference attributes requires the consideration of dominance relations in data analysis. Object x dominates object y if the following holds:

1. Object x is at least as good as object y in all considered preference attributes.
2. Object x and y have identical or similar description in all considered preference attributes.

According to rational behavior, assigning objects to preference information classes should meet the following dominance principle: if object x dominates object y, then x should be assigned to a class no worse than y. When objects x and y do not satisfy the dominance principle, the pair (x, y) is inconsistent. Decision classes can express preference. For example, the decision-maker could divide bankruptcy risk into three categories: high risk, medium risk, and low risk. Such classification is also called order sorting.

4.1.2 Dominating Sets and Dominated Sets

To deal with the ordinal property of the preference attributes in rough approximation, the indiscernibility relation should be replaced by the dominance relation.

Definition 4.1 Suppose $x, y \in U$, $P \subseteq C$, for $\forall q \in P$, if $f(y, q) \geq f(x, q)$, then y dominates x with respect to $P \subseteq C$, denoted by yD_Px. D_P is called dominance relation.

Dominance relation defined above is actually a weak preference relation on U expressing a preference on the set of objects with respect to a preference attribute q. If $f(y, q) = f(x, q)$ holds for $\forall q \in P$, dominance relation comes back to indiscernibility relation.

Proposition 4.1 Dominance relation is reflexive and transitive.

Proof: For $\forall x \in U$, then xD_Px, it can be inferred that dominating reference is reflexive. For $\forall x, y, z \in U$, if $f(y, q) \geq f(x, q)$ and $f(x, q) \geq f(z, q)$, then $f(y, q) \geq f(z, q)$, so dominance relation is transitive.

Definition 4.2 Suppose $P \subseteq C$ and $x \in U, y \in U$, P-dominating set and P-dominated set with respect to x, are defined, respectively, as

$$D_P^+(x) = \{yD_Px\}$$

$$D_P^-(x) = \{xD_Py\}$$

Suppose the object set U can be divided into the finite number of decision classes according to a decision-maker, while $Cl = \{Cl_t, t \in \{1, 2, \ldots, n\}\}$ is a set of classes of U, then we can obtain the following upward unions and downward unions of the classes with $Cl_n > \cdots > Cl_t > \cdots > Cl_1$:

$$Cl_t^{\geq} = \bigcup_{s \geq t} Cl_s, \quad Cl_t^{\leq} = \bigcup_{s \leq t} Cl_s \quad t, s \in \{1, 2, \ldots, n\}$$

The statement $x \in Cl_t^{\geq}$ implies that x at least belongs to class Cl_t or a more preferred class, while $x \in Cl_t^{\leq}$ implies that x at most belongs to class Cl_t or a less preferred class. Obviously, each object from the upward union $Cl_t^{\geq}$ is at least as good as each object from the downward union $Cl_t^{\leq}$, $Cl_1^{\geq} = Cl_n^{\leq} = U$ and $Cl_n^{\geq} = Cl_n$. Moreover, for $t = 2, \ldots, n$, then $Cl_{t-1}^{\leq} = U - Cl_t^{\geq}$ and $Cl_t^{\geq} = U - Cl_{t-1}^{\leq}$.

In other words, the decision-maker has assigned the objects in the universe U to the classes Cl according to a comprehensive evaluation: the worst objects are in Cl_1, while the best objects are in Cl_n, and the other objects belong to the remaining classes Cl_r. For all $r, t \in \{1, \ldots, n\}$ such that $r > t$, the objects from Cl_r are preferred to the objects from Cl_t.

4.1.3 Rough Approximation by Means of Dominance Relations

Suppose $S = (U, A, V, f)$ is a preference attribute decision table, $A = C \cup D$, and given attribute sets with preference information $P \subseteq C$, $Cl_t \subseteq D$, $x \in U$, $t \in \{1, \ldots, n\}$. The P-lower and the P-upper approximations of $Cl_t^{\geq}$, proposed according to Greco's theory, are defined, respectively, as follows:

$$\underline{apr}_P(Cl_t^{\geq}) = \bigcup\{x \in U : D_P^+(x) \subseteq Cl_t^{\geq}\} \tag{4.1}$$

$$\overline{apr}_P(Cl_t^{\geq}) = \bigcup\{x \in U : D_P^-(x) \cap Cl_t^{\geq} \neq \phi\} \tag{4.2}$$

The P-lower approximation of $Cl_t^{\geq}$ $\underline{apr}_P(Cl_t^{\geq})$ can be interpreted as the union of all condition classes which have the property that the objects are certainly classified to the decision class $Cl_t^{\geq}$, while the P-upper approximation of $Cl_t^{\geq}$ $\overline{apr}_P(Cl_t^{\geq})$ can be interpreted as the union of all classes which have the property that the objects are possibly classified to the decision class $Cl_t^{\geq}$.

The boundary of $Cl_t^{\geq}$ is defined as

$$bnd_P(Cl_t^{\geq}) = \overline{apr}_P(Cl_t^{\geq}) - \underline{apr}_P(Cl_t^{\geq}) \tag{4.3}$$

Analogously, the P-lower and the P-upper approximations of $Cl_t^{\leq}$, respectively, are defined as follows:

$$\underline{apr}_P(Cl_t^{\leq}) = \bigcup\{x \in U : D_P^-(x) \subseteq Cl_t^{\leq}\} \tag{4.4}$$

$$\overline{apr}_P(Cl_t^{\leq}) = \bigcup\{x \in U : D_P^+(x) \cap Cl_t^{\leq} \neq \phi\} \tag{4.5}$$

The P-lower approximation of $Cl_t^{\leq}$ $\underline{apr}_P(Cl_t^{\leq})$ can be interpreted as the union of all condition classes that have the property that objects are certainly classified to the decision class $Cl_t^{\leq}$, while the P-upper approximations of $Cl_t^{\leq}$ $\overline{apr}_P(Cl_t^{\leq})$ can be interpreted as the union of all condition classes that have the property that objects are possibly classified to the decision class $Cl_t^{\leq}$.

The boundary of $Cl_t^{\leq}$ is defined as

$$bnd_P(Cl_t^{\leq}) = \overline{apr}_P(Cl_t^{\leq}) - \underline{apr}_P(Cl_t^{\leq}) \tag{4.6}$$

Elements in boundary of $Cl_t^{\leq}$ or $Cl_t^{\geq}$ cannot illustrate their belonging definitely.

4.1.4 Classification Quality and Reduct

The measure of classification quality is defined as

$$\gamma_P(Cl) = \frac{|U - ((\bigcup bnd(Cl_t^{\geq})) \cup (\bigcup bnd(Cl_t^{\leq})))|}{|U|} \tag{4.7}$$

Classification quality $\gamma_P(Cl)$ expresses the ratio of the relation between all the P-correctly classified objects and all the objects in the multicriteria decision table. For every minimal subset $P \subseteq C$, $\gamma_P(Cl) = \gamma_C(Cl)$ is called a reduct of C with respect to Cl and is denoted by $red_{Cl}(P)$.

4.1.5 Preferential Decision Rules

The preferential decision rule is one kind of dependence form between condition preference attribute and decision preference attribute. After obtaining rough approximation based on dominance relation, the following preferential decision rules can be induced through preferential information in the decision table.

$D_{\geq}$-decision rules, which have the following form:

$$\text{if } f(x, q_1) \geq r_{q1} \wedge f(x, q_2) \geq r_{q2} \wedge \cdots \wedge f(x, q_P) \geq r_{qp}, \text{ then } x \in Cl_t^{\geq}$$

$D_{\leq}$-decision rules, which have the following form:

$$\text{if } f(x, q_1) \leq r_{q1} \wedge f(x, q_2) \leq r_{q2} \wedge \cdots \wedge f(x, q_P) \leq r_{qp}, \text{ then } x \in Cl_t^{\leq}$$

where $\{q_1, q_2, \ldots, q_p\} \subseteq C$, $(r_{q1}, r_{q2}, \ldots, r_{qp}\} \in V_{q1} \times V_{q2} \times \cdots \times V_{qp}$, $t \in \{1, 2, \ldots, n\}$.

Example 4.1 A middle school principal wants to assign students to two classes according to their scores: bad and good. To obtain classification rules, the principal collects six students' scores of three subjects in one class, and they are shown in Table 4.1, while the meanings of the attributes in the table are as follows: a_1 represents grade in mathematics, a_2 represents grade in physics, and a_3 represents grade in literature, while d represents comprehensive evaluation.

1. According to the dominance-based RST, calculate lower approximation and upper approximation, classification quality, and reduct, then generate preferential decision rules based on deduct.
2. Compare and analyze the calculation results of the dominance-based rough sets theory and RST.

Solution: According to Table 4.1, $S = (U, A, V, f)$ is an information system, where the universe $U = \{n_1, n_2, n_3, n_4, n_5, n_6\}$, condition attribute set $C = \{a_1, a_2, a_3\}$, and decision attribute set $D = \{d\}$, obviously, a_1, a_2, a_3, d are preference attributes. For a_1, good > medium > bad; for a_2, good > bad; for a_3, good > bad; and for d, good > bad.

Table 4.1 Evaluation Table of Student's Examination Scores

U	Condition Attributes (C)			Decision Attribute (D)d
	a_1	a_2	a_3	
n_1	Good	Good	Bad	Good
n_2	Medium	Bad	Bad	Bad
n_3	Medium	Bad	Bad	Good
n_4	Bad	Bad	Bad	Bad
n_5	Medium	Good	Good	Bad
n_6	Good	Bad	Good	Good

Suppose $Cl_1^{\leq}$ is at most bad students; $Cl_2^{\geq}$ is at least good students. Because only two classes are considered, we have $Cl_1^{\leq} = Cl_1(bad)$, $Cl_2^{\geq} = Cl_2(good)$.

Based on dominance relations, we can divide the universe according to decision classification. The following classifications can be concluded:

$$\frac{U}{C} = \{X_1, X_2, X_3, X_4\}$$

where $X_1 = \{n_1\}$, $X_2 = \{n_2, n_3, n_5\}$, $X_3 = \{n_4\}$, $X_4 = \{n_6\}$;

$$\frac{U}{D} = \{Cl_1, Cl_2\}$$

where $Cl_1 = \{n_2, n_4, n_5\}$, $Cl_2 = \{n_1, n_3, n_6\}$.

Then the lower approximation of $Cl_1^{\leq}$ $\underline{apr}_P(Cl_1^{\leq})$ and the upper approximation $\overline{apr}_P(Cl_1^{\leq})$, the lower approximation of $Cl_2^{\geq}$ $\underline{apr}_P(Cl_2^{\geq})$ and the upper approximation $\overline{apr}_P(Cl_2^{\geq})$, respectively, are as follows:

$$\underline{apr}_P(Cl_1^{\leq}) = \{n_4\}$$

$$\overline{apr}_P(Cl_1^{\leq}) = \{\{n_2, n_3, n_5\}, \{n_4\}\}$$

$$\underline{apr}_P(Cl_2^{\geq}) = \{\{n_1\}, \{n_6\}\}$$

$$\overline{apr}_P(Cl_2^{\geq}) = \{\{n_1\}, \{n_6\}, \{n_2, n_3, n_5\}\}$$

$$\gamma_P(Cl) = \frac{|\{n_4\} \cup \{n_1, n_6\}|}{|\{n_1, n_2, n_3, n_4, n_5, n_6\}|} = \frac{3}{6} = 0.5$$

According to the definition of the dominance-based rough sets theory, the minimum reduct $\{a_1\}$ can be obtained; obviously it is also a core. Thus, we can obtain the minimum preferential decision rule sets shown in Table 4.2.

Table 4.2 Minimum Preferential Decision Rules Set Generated by Reduct $\{a_1\}$

Rules	*Confidence of the Support Numbers*
If $a_1 \geq$ good, then $d \in Cl_2$	2
If $a_1 \leq$ bad, then $d \in Cl_1$	1

Table 4.3 Minimum Decision Rule Sets Generated by Reduct $\{a_1, a_2\}$

Rules	*Support Numbers*
If a_1 = good, then $d \in Cl_2$	2
If a_1 = medium and a_2 = good, then $d \in Cl_1$	1
If a_1 = bad, then $d \in Cl_1$	1

According to RST, one minimum reduct acquired is $\{a_1, a_2\}$, thus we can acquire the minimum decision rule sets shown in Table 4.3 based on the reduct $\{a_1, a_2\}$.

According to dominance relation, student 5 is better than student 3, because for all the three preference condition attributes, student 5 is at least as well as student 3; however, the comprehensive evaluation of student 5 is worse than that of student 3. The inconsistency can be approximately explained as incompatibility based on dominance relation, but it cannot be approximately identified based on indiscernibility relation. In the sense of indiscernibility relation, student 5 and student 3 is discernible.

4.2 Dominance-Based Variable Precision Rough Set

Similar to rough set model, dominance-based rough set model cannot obtain probabilistic decision rule under the condition that noise data exists. In the real world, noise data is unavoidable. To find probability rule from multicriteria decision table, the indiscernibility relation can be replaced by dominance relation and the confident threshold value is given; thus, rough set model could be extended to the approach that can obtain decision rule from multicriteria decision table, so that preference model can be inducted by an exemplar decision system that contains noise data according to probabilistic decision rule.

4.2.1 Inconsistency and Indiscernibility Based on Dominance Relation

Given a subset of preference attributes $P \subseteq C$, $x \in U$, $y \in U$, for the inclusion of $x \in U$ to the upward union of classes $Cl_t^{\geq}$, where $t \in \{1, 2, \ldots, n\}$, an inconsistency based on the dominance relation is created if one of the following conditions holds:

1. $x \in Cl_t^{\geq}$, that is, x belongs to Cl_t or better than Cl_t, if in all the preference attributes considered, y is at least as good as x, but y belongs to a worse class than x. That is $x \in Cl_t^{\geq}$, but $D_P^+(x) \cap Cl_{t-1}^{\leq} \neq \phi$.
2. $x \notin Cl_t^{\geq}$, that is, x belongs to classification worse than Cl_t, if in all the preference attributes considered, x is at least as good as y, but y belongs to a better class than x. That is, $x \notin Cl_t^{\geq}$, but $D_P^-(x) \cap Cl_t^{\geq} \neq \phi$.

When preference information of Cl_t is not considered, that is $Cl_t^{\geq} = Cl_t^{\leq}$, inconsistencies in the sense of dominance relation is degenerated to indiscernibility in the sense of indiscernibility relation.

Obviously, discernible classes based on dominance relation include the following: condition classes are indiscernible classes based on equivalence relation; according to decision preference attribute, condition classes are inconsistence classes based on dominance relation. If the inconsistence is rather small, such as confidence more than or equal to a given threshold value, then probabilistic decision rule of preference model can be obtained according to variable precision rough set model.

4.2.2 β-Rough Approximation Based on Dominance Relations

Definition 4.3 Suppose information system $S = (U, A, V, f)$ is a preference attribute decision table, where $A = C \cup D$, and given a preference attribute set $P \subseteq C$, $X \subseteq U$, $Cl_t \subseteq D$, $t \in \{1, \ldots, n\}$, confidence threshold value $0.5 < \beta \leq 1$, then the β-lower and β-upper approximations of $Cl_t^{\geq}$ are respectively defined as follows:

$$\underline{apr}_P^{\beta}(Cl_t^{\geq}) = \bigcup \left\{ \frac{|D_P^+(x) \cap Cl_t^{\geq}|}{|D_P^+(x)|} \geq \beta \right\} \tag{4.8}$$

$$\overline{apr}_P^{\beta}(Cl_t^{\geq}) = \bigcup \left\{ \frac{|D_P^-(x) \cap Cl_t^{\geq}|}{|D_P^-(x)|} > 1 - \beta \right\} \tag{4.9}$$

The β-lower approximation of $Cl_t^{\geq}$ can be interpreted as the union of all condition classes with confidence threshold value greater than or equal to β that have the property that the objects certainly belong to the upward union $Cl_t^{\geq}$, while the β-upper approximations of $Cl_t^{\geq}$ can be interpreted as the union of all classes that have the property that the objects possibly belong to the upward union $Cl_t^{\geq}$ and the probability greater than 1 – β.

Analogously, the β-lower and β-upper approximations of $Cl_t^{\leq}$, respectively, are defined as follows:

$$\underline{apr}_P^{\beta}(Cl_t^{\leq}) = \bigcup \left\{ \frac{|D_P^-(x) \cap Cl_t^{\leq}|}{|D_P^-(x)|} \geq \beta \right\} \tag{4.10}$$

$$\overline{apr}_P^{\beta}(Cl_t^{\leq}) = \bigcup \left\{ \frac{|D_P^+(x) \cap Cl_t^{\leq}|}{|D_P^+(x)|} > 1 - \beta \right\} \tag{4.11}$$

The β-lower approximation of $Cl_t^{\le}$ can be interpreted as the union of all condition classes with confidence threshold value greater than or equal to β which have the property that objects certainly belong to the downward union $Cl_t^{\le}$, while the β-upper approximations of $Cl_t^{\le}$ can be interpreted as the union of all classes which have the property that objects possibly belong to the downward union $Cl_t^{\le}$ and the probability greater than 1 – β. Such defined approximation takes both the discernibility based on equivalence relation and inconsistence based on dominance relation into consideration.

Proposition 4.2 For $\forall P \subseteq C$, $Cl_t \subseteq D$, $\forall t \in \{1, 2, \ldots, n\}$ and $0.5 < \beta \le 1$, then

1. $\underline{apr}_P^{\beta}(Cl_t^{\ge}) \subseteq \overline{apr}_P^{\beta}(Cl_t^{\ge})$
2. $\underline{apr}_P^{\beta}(Cl_t^{\le}) \subseteq \overline{apr}_P^{\beta}(Cl_t^{\le})$

Proof: For $x \in \underline{apr}_P^{\beta}(Cl_t^{\ge})$, because D_P is reflexive, then $x \in D_P^{+}(x)$, according to $0.5 < \beta \le 1$, then $\beta > 1 - \beta$, for D_P is reflexive, then $x \in D_P^{-}(x)$, and we can induce from this $x \in \overline{apr}_P^{\beta}(Cl_t^{\ge})$, so $\underline{apr}_P^{\beta}(Cl_t^{\ge}) \subseteq \overline{apr}_P^{\beta}(Cl_t^{\ge})$. $\underline{apr}_P^{\beta}(Cl_t^{\le}) \subseteq \overline{apr}_P^{\beta}(Cl_t^{\le})$ can be demonstrated similarly.

Proposition 4.3 When β = 1, the β-lower approximation and β-upper approximation of $Cl_t^{\ge}$ and the β-lower and β-upper approximations of $Cl_t^{\le}$ are degenerated to dominance rough approximation.

Proof: When β = 1, $(|D_P^{+}(x) \cap Cl_t^{\ge}| / |D_P^{+}(x)|) \ge \beta = 1$, then $D_P^{+} \subseteq Cl_t^{\ge}$; according to $(|D_P^{-}(x) \cap Cl_t^{\le}| / |D_P^{-}(x)|) \ge \beta = 1$, then $D_P^{-} \subseteq Cl_t^{\le}$, so the β-lower approximation of $Cl_t^{\ge}$ $\underline{apr}_P^{\beta}(Cl_t^{\ge})$ and the β-lower approximation of $Cl_t^{\le}$ $\underline{apr}_P^{\beta}(Cl_t^{\le})$ are lower approximation of dominance sets, if $(|D_P^{-}(x) \cap Cl_t^{\ge}| / |D_P^{-}(x)|) > 1 - \beta = 0$, then $D_P^{-}(x) \cap Cl^{\ge} \ne \phi$, $(|D_P^{+}(x) \cap Cl_t^{\le}| / |D_P^{+}(x)|) > 1 - \beta = 0$, and $D_P^{+}(x) \cap Cl^{\le} \ne \phi$, so β-upper approximation of $Cl_t^{\ge}$ $\overline{apr}_P^{\beta}(Cl_t^{\ge})$ and β-upper approximation of $Cl_t^{\le}$ $\overline{apr}_P^{\beta}(Cl_t^{\le})$ are *P*-upper approximation of dominance sets.

Proposition 4.4 When the preferential information in decision table is not considered, the proposed approach is degenerated to variable precision rough set approach. When β = 1 and preference information in decision table is not considered, the proposed approach is degenerated to rough set approach.

Proof: When the preference information in decision table is not considered, dominance relation is degenerated to indiscernibility relation, and $Cl_t^{\ge} = Cl_t^{\le}$. Obviously, the proposed model is degenerated to variable precision rough set model. When β = 1, variable precision rough set model is rough set model.

4.2.3 Classification Quality and Approximate Reduct

Proposition 4.5 Classification quality of Cl is defined as

$$\gamma_P^\beta(Cl) = \frac{\left|\left(\bigcup\left\{\frac{|D_P^+(x)\cap Cl_t^{\geq}|}{|D_P^+(x)|} \geq \beta\right\}\right) \cup \left(\bigcup\left\{\frac{|D_P^-(x)\cap Cl_t^{\leq}|}{|D_P^-(x)|} \geq \beta\right\}\right)\right|}{|U|} \tag{4.12}$$

Classification quality $\gamma_P^\beta(Cl)$ measures the proportion of objects in preferential decision table for which a classification is possible at the specified confident threshold value β in universe.

The approximate reduct $red_P^\beta(C, Cl)$ is defined as the minimal subset $P \subseteq C$ that satisfies $\gamma_P^\beta(Cl) = \gamma_C^\beta(Cl)$ for a specified threshold value β. There is a possibility of more than one reduct in a decision table, and the intersection of all reducts is called the core.

4.2.4 Preferential Probabilistic Decision Rules

For preference information system $S = (U, A, V, f)$, $A = C \cup D$, condition classes in S can be denoted by $X_i (i = 1, 2, \ldots, k)$; decision classes in S can be denoted by Cl_t $(t = 1, 2, \ldots, n)$, and $X_i \cap Cl_t = \phi$. The preferential probabilistic decision rule sets are divided into two classes: the upward union $D_{\geq}$ of preferential probabilistic decision rules and the downward union $D_{\leq}$ of preferential probabilistic decision rules, then the rule

$$r\colon\ CON_C(X_i) \xrightarrow{\beta} DEC_D(Cl_t) \tag{4.13}$$

is called the preferential probabilistic decision rules of (C, D), denoted by $\{r_{it}\}$ and confidence is β.

The form of $D_{\geq}$ preferential probabilistic decision rules is as follows:

If $f(x, q_1) \geq r_{q1} \wedge f(x, q_2) \geq r_{q2} \ldots \wedge f(x, q_p) \geq r_{pq}$, then $x \in Cl_t^{\geq}$, with confidence β, where $\{q_1, q_2, \ldots, q_p\} \subseteq C$, $(r_{q1}, r_{q2}, \ldots, r_{qp}) \in V_{q1} \times V_{q2} \times \cdots \times V_{qp}$. These rules are supported by objects of the β-lower approximation of an upward union $Cl_t^{\geq}$.

The form of $D_{\leq}$ preferential probabilistic decision rules is as follows:

If $\{q_1, q_2, \ldots, q_p\} \subseteq C$, $(r_{q1}, r_{q2}, \ldots, r_{qp}) \in V_{q1} \times V_{q2} \times \cdots \times V_{qp}$, then $x \in Cl_t^{\leq}$ with confidence β.

These rules are supported by objects of the β-lower approximation of a downward union $Cl_t^{\leq}$.

4.2.5 Algorithm Design

In this section, a new algorithm on the basis of the Apriori algorithm is introduced to mine preference multiattribute probability rules. The main idea is as follows: for each preference rule, check whether it is greater than or equal to a given confidence threshold β, then put the rule that meets the requirements as knowledge into rule base. The Apriori algorithm uses an iterative approach that uses k-itemset to find (k + l)-itemset. First of all, the frequency 1-itemsets must be found and denoted by L_1, then frequency 2-itemsets L_2 are generated from L_1, and L_3 are generated from L_2, and so on until frequency k-itemsets cannot be found. The specific algorithm is described as follows:

Input: Preferential multiattribute decision table with confidence β.
Output: Preferential probabilistic decision rules.
Step 1: Calculate indiscernibility based on dominance relations of each of the preference attribute value for $[f(x,q_i) = r_{qi}]$, and form one variable candidate set S_1.
Step 2: Suppose $L_1 = S_1$, $k = 1$, $m = 1$.
Step 3: If S_k is nonempty, the $FD = |D_P^+(x) \cap Cl_t^{\geq}|/|D_P^+(x)|$ for each candidate pattern $\wedge_k[f(x,q_k) = r_{qk}]$ in the candidate set S_k can be calculated, the pattern of $FD > 0$ can be saved into the set RS_k.
Step 4: If L_m is nonempty, the $LD = |D_P^-(x) \cap Cl_t^{\leq}|/|D_P^-(x)|$ for each candidate pattern $\wedge_m[f(x,q_m) = r_{qm}]$ in the candidate set L_m can be counted, the pattern of $LD > 0$ can be saved into the set RL_m.
Step 5: For each candidate pattern $\wedge_k[f(x,q_k) = r_{qk}]$ in RS_k, if $FD > \beta$, then this pattern can be saved in the rule repository, at the same time it should be deleted from RS_k.
Step 6: For each candidate pattern $\wedge_m[f(x,q_m) = r_{qm}]$ in RL_m, if $LD > \beta$, then this pattern can be saved in the rule base; meanwhile, it should be deleted from RL_m.
Step 7: If RS_k is nonempty, then intersection operation of sets pattern $\wedge_k[f(x,q_k) = r_{qk}]$ in RS_k can generate S_{k+1}.
Step 8: If RL_m is nonempty, then intersection operation of sets pattern $\wedge_m[f(x,q_m) = r_{qm}]$ in RL_m can generate L_{m+1}.
Step 9: If the pattern $\wedge_k[f(x_i,q) = r_{qj}]$ is not in RS_k, then deletion calculation should be done in S_{k+1}.
Step 10: If the pattern $\wedge_m[f(x_i,q) = r_{qj}]$ is not in RL_m, then deletion calculation should be done in L_{m+1}.
Step 11: Suppose $k = k + 1$, $m = m + 1$.
Step 12: Repeat Step 3 to Step 11, until the β-reduct is found, or RS_k and RL_m is empty set.

Example 4.2 A consulting agency has evaluated the bankruptcy risk situation of eight banks, and the knowledge representation system of corporate bankruptcy risk assessment is shown in Table 4.4.

In Table 4.4, a_1: profit rate, that is, the rate of operating profit and operating income of a company.

Operating profit ratio reflects the proportion of operating profit of operating revenues. For investors, the higher the ratio is, the better. The higher operating profit ratio shows higher profit levels in business operating revenues.

a_2: return on capital, that is, the rate of the total corporate profits to the capital corporate business owners invest.

Table 4.4 Knowledge Representation System of Corporate Bankruptcy Risk Assessment

U	Condition (*C*)			*Bankruptcy Risk (D) d*
	a_1	a_2	a_3	
n_1	High	High	High	Low risk
n_2	Medium	Low	High	High risk
n_3	Medium	Low	High	High risk
n_4	Medium	Low	High	Low risk
n_5	Low	Low	High	High risk
n_6	Medium	High	Low	High risk
n_7	High	Low	Low	Low risk
n_8	Medium	Low	High	High risk

Return on capital reflects the capital profitability. In view of the investors, the higher the return on capital is, the better. Higher return on capital means that investors gain more profits after investing the same capital. The profitability of the company is stronger.

a_3: asset–liability ratio, also known as debt ratio, is the ratio of the company's total liabilities (including current liabilities and long-term liabilities) and total assets.

Asset–liability ratio reflects how much proportion of the company's total assets is gained through borrowing loans from the creditor. The ratio reflects the company's debt situation, debt capacity, and the degree of creditor protection. For creditors, the lower the ratio of liabilities is, the better, because the low debt ratio shows that the company's total liabilities in the proportion of the assets are smaller. The higher level of protection means the risk is smaller; on the contrary, the higher debt ratio shows the proportion of the debt in the company's total assets is greater, and the lower the level of protection is. When the economy deteriorates, the company's debt burden is on the increase, the occurrence of losses and bankruptcy more possible, thus the risk of claims greater.

Please try to judge for the corporate bankruptcy in Table 4.4.

Solution: In Table 4.4, condition attribute set $C = \{a_1, a_2, a_3\}$, and decision attribute set $D = \{d\}$, obviously, a_1, a_2, a_3, d are standards. For a_1, high > middle > low; for a_2, high > low; for a_3, low > high; and for d, low risk > high risk.

Suppose $Cl_1^{\leq}$ is at most high risk enterprise; $Cl_2^{\geq}$ is at least low risk enterprise, only considering two kinds, so $Cl_1^{\leq} = Cl_1(\textit{high risk})$, $Cl_2^{\geq} = Cl_2(\textit{low risk})$.

Divide the universe according to the decision classifications and the following classifications can be acquired based on dominance relation:

$$U/C = \{X_1, X_2, X_3, X_4\}$$

where $X_1 = \{n_1\}$, $X_2 = \{n_2, n_3, n_4, n_6, n_8\}$, $X_3 = \{n_5\}$, $X_4 = \{n_7\}$;

$$U/D = \{Cl_1, Cl_2\}$$

where $Cl_1 = \{n_2, n_3, n_5, n_6, n_8\}$, $Cl_2 = \{n_1, n_5, n_7\}$.

Suppose β = 75 percent, then the β-lower approximation $\overline{apr}_P^{\beta}(Cl_1^{\leq})$ and the β-upper approximation $\overline{apr}_P^{\beta}(Cl_1^{\leq})$ of $Cl_1^{\leq}$, and the β-lower approximation $\underline{apr}_P^{\beta}(Cl_2^{\geq})$ and the β-upper approximation $\overline{apr}_P^{\beta}(Cl_2^{\geq})$ of $Cl_2^{\geq}$ are described, respectively as, follows:

$$\underline{apr}_P^{\beta}(Cl_1^{\leq}) = \{\{n_2, n_3, n_4, n_6, n_8\}, \{n_5\}\}$$

$$\overline{apr}_P^{\beta}(Cl_1^{\leq}) = \{\{n_2, n_3, n_4, n_6, n_8\}, \{n_5\}\}$$

$$\underline{apr}_P^{\beta}(Cl_2^{\geq}) = \{\{n_1\}, \{n_7\}\}$$

$$\overline{apr}_P^{\beta}(Cl_2^{\geq}) = \{\{n_1\}, \{n_7\}\}$$

According to the algorithm in Section 4.2.5, a minimal β-reduct $\{a_1\}$ can be acquired; obviously, the reduct is a core. Thus, we can acquire the minimal probabilistic decision rule sets shown in Table 4.5.

From Table 4.5, we can know the classification quality $\gamma_P^{\beta}(Cl) = 1$.

We can obtain the minimal attribute reduct $\{a_1, a_2\}$ with variable precision rough set based on indiscernibility relation, thus the minimal probabilistic decision rule sets can be acquired with the reduct $\{a_1, a_2\}$ and shown in Table 4.6.

According to dominance relation, enterprise 6 is better than enterprise 4, because as far as all the three standards are concerned, enterprise 6 is at least as good as enterprise 4. But the comprehensive evaluation of enterprise 6 is worse than enterprise 4. Therefore, it is approximately explained as inconsistence based on dominance relation, but it cannot get approximate identification in the light of indiscernibility relation. In the sense of indiscernibility relation, enterprise 6 and enterprise 4 are discernible. The number of conditions used to generate probabilistic decision rules in Table 4.5 is smaller than that in Table 4.6, meanwhile the number of rules inducted is smaller but stronger.

Table 4.5 A Minimal Probabilistic Decision Rule Set

Rules	*Support Numbers*	*Confidence Level (Percent)*
$a_1 \leq \text{medium} \xrightarrow{\text{80 percent}} d \in Cl_1$	6	80
$a_1 \geq \text{high} \xrightarrow{\text{100 percent}} d \in Cl_2$	2	100

Table 4.6 Minimal Probabilistic Decision Rule Sets Generated from β-Reduct $\{a_1, a_2\}$

Rules	*Support Counts*	*Confidence Level (Percent)*
$a_1 = \text{high} \xrightarrow{100\text{ percent}} d \in Cl_2$	2	100
$a_1 = \text{medium} \wedge a_2 \xrightarrow{75\text{ percent}} d \in Cl_1$	4	75
$a_1 = \text{medium} \wedge = \text{high} \xrightarrow{100\text{ percent}} d \in Cl_1$	1	100
$a_1 = \text{low} \xrightarrow{100\text{ percent}} d \in Cl_1$	1	100

4.3 An Application Case

4.3.1 Post-Evaluation of Construction Projects Based on Dominance-Based Rough Set

Post-evaluation of construction project is to totally, systematically, and objectively analyze the aim, process, benefit, function, and influence after having finished a construction project. By checking and summing up the practice of construction project, it is made sure whether the expectative aim has been attained, whether the project is reasonable and effective, whether the main benefit has been realized. By means of evaluation, the reason for success or failure can be discovered; experience and lesson can be summarized. Further, the suggestion on investment decision of future projects is provided, and improved suggestion is proposed for the problems of project implementation and management through timely valid information feedback, so that the legalization management of construction project investment is strengthened, and the aim of increasing economic benefit can be reached. The post-evaluation of a construction project can be divided into such categories as process evaluation, benefit evaluation, influence evaluation, and continuable evaluation. Research on post-evaluation of construction projects are now still at their initial stage in China, and have not yet formed scientific approach and system.

For indexes of the post-evaluation of a construction project, such as quality evaluation, prophase management evaluation, process management evaluation, and comprehensive evaluation, we always expect score values to be higher. This means that these condition and decision indexes contain preference information. Dominance-based rough set analysis is used in the post-evaluation of construction projects to explore a scientific normative evaluation approach so that we can make reasonable classification and classification evaluation on the post-evaluation projects, and can thus meet the appraisal requirements of the post-evaluation projects and realize the aim of summarization and evaluation. It can provide some scientific suggestions on improvement of management and decision of investment projects.

4.3.1.1 Construction of Preferential Evaluation Decision Table

The process evaluation in the post-evaluation of a construction project is to differentiate reasonable degree of the deviation between the real result and expected result, make analysis of the reason for inducing deviation, and then find and solve the existing problems in decision and implementation in time on the basis of the feasible study of the construction project by following the policy and laws of the national promulgation, approach, and regulation, and so on, set by investors or relevant management section from the perspectives of the construction project itself, national and social economy, so that the practical and feasible measures and suggestions can be proposed to improve operation. At the same time, experiences and lessons can be drawn to help to accumulate experiences for both strengthening former work and enhancing project management. In addition, it can provide valuable data for the follow-up project management. The process evaluation should cover each stage of the construction project and be able to reflect the characteristics of main links at each stage.

There are 10 first-level indexes and 58 second-level indexes in the index system of process evaluation in Table 4.7. They are used to determine the weight of each index based on the "Delphi" approach, then these indexes and their weights are shown in Table 4.7.

According to the 58 scores of second-level indexes of the project and their corresponding weights given by experts, we, first of all, calculate the scores of 10 first-level indexes, and then calculate the synthesis score of this project based on the scores of the first-level and their weights.

For example, the first-level index of the prophase management evaluation contains five second-level indexes, which are project preparation, project proposal, feasibility study report, preliminary design scheme, and bidding management. Their weight vectors are as follows:

$$\sigma_1 = (0.12, 0.14, 0.29, 0.27, 0.18\}$$

If their marks vectors are as follows,

$$x_1 = (97, 98, 97, 94, 97)$$

then first-level index of the prophase management evaluation gets scores as follows:

$$G_1 = \sigma_1 x_1^T = (0.12, 0.14, 0.29, 0.27, 0.18) \cdot (97, 98, 97, 94, 97)^T = 96.33$$

Similarly, we can obtain the comprehensive score of the project according to scores of the first-level indices and their corresponding weights. Several experts will be invited to evaluate a certain project to give scores, and the average score of the comprehensive scores will be taken as the final score. The decision table of process evaluation of constructive projects is shown in Table 4.8 according to the scores of the 18 post-evaluation projects from experts.

Table 4.7 Index System and Weights of the Process Evaluation of Construction Projects

First-Level Indices and Weights	*Second-Level Indices*	*Weights*
Evaluation of prophase management (0.09)	Project preparation	0.12
	Project proposal book	0.14
	Feasibility research report	0.29
	Preliminary design scheme	0.27
	Bidding management	0.18
Evaluation of construction period management (0.11)	Organization management	0.12
	Supervision system	0.13
	Contract management	0.15
	Financc management	0.14
	Stage rate of completion	0.10
	Design schedule	0.10
	Schedule control	0.12
	Management system	0.14
Evaluation of trial production (0.09)	Production arrangement	0.13
	Personnel training	0.20
	Preparation for trial production	0.17
	Results of trial production	0.28
	Technical check	0.22
Evaluation of technology (0.10)	Technological process	0.36
	Technological route	0.31
	Technological innovation	0.33
Evaluation of fire fighting system (0.08)	Building fire protection	0.18
	Fire-control system	0.17
	Fire-control equipments	0.13

(continued)

Table 4.7 (continued) Index System and Weights of the Process Evaluation of Construction Projects

First-Level Indices and Weights	*Second-Level Indices*	*Weights*
	Training of firemen	0.14
	Fire-control management	0.14
	Accident emergency	0.11
	Alarm system	0.13
Evaluation of project construction quality (0.20)	Impressions	0.10
	Technical data	0.10
	Quality of pile foundation project	0.15
	Quality of foundation project	0.18
	Quality of main body project	0.22
	Quality of single project	0.14
	Accidents of project quality	0.11
Evaluation of hygienic safety (0.07)	Spot-cleaning	0.18
	Hygienic safety	0.25
	Safety measures	0.30
	Three simultaneousness of hygiene safety	0.27
Evaluation of environmental protection (0.10)	Approval process	0.09
	Three simultaneousness of environmental protection	0.12
	Environmental protection facilities	0.11
	Running status of environmental protection facilities	0.13
	Discharge of three main kinds of wastes	0.16

Table 4.7 (continued) Index System and Weights of the Process Evaluation of Construction Projects

First-Level Indices and Weights	*Second-Level Indices*	*Weights*
	Ecological protection	0.11
	Personnel allocation	0.08
	Clean production	0.08
	Measures of pollution control	0.13
Evaluation of archives (0.05)	Filing efficiency of dossiers	0.20
	Accuracy of dossiers	0.30
	Systematization of dossiers	0.28
	Dossiers management	0.22
Evaluation of design and budgetary estimate control (0.11)	General plan layout	0.18
	Level of design	0.23
	Control of design change	0.13
	Enforcement of budgetary estimate	0.18
	Arrival rate of fund	0.15
	Examine reduction	0.13

In Table 4.8, conditional attribute set $C = \{a_1, a_2, a_3, a_4, a_5, a_6, a_7, a_8, a_9, a_{10}\}$, decision attribute set $D = \{d\}$, where a_1 stands for evaluation of prophase management; a_2 stands for evaluation of construction period management; a_3 is evaluation of trail production; a_4 is evaluation of technology; a_5 is evaluation of fire fighting system; a_6 is evaluation of project construction quality; a_7 is evaluation of the hygienic safety; a_8 is evaluation of environmental protection; a_9 is evaluation of archives; a_{10} is evaluation of design and budgetary estimate; d is comprehensive evaluation.

According to the grade criteria of post-evaluation for construction projects, each index (including comprehensive evaluation) can be divided into four grades: excellent, good, eligible, and bad, among which "excellent" means that this index gets a score between 90 and 100, containing 90; "good" means that it gets a score between 80 and 90, containing 80; "eligible" means that it gets a score between 70 and 80, containing 70; and "bad" index means that it get a score under 70. Based on the above grade classification, the data in Table 4.8 is discredited into a preferential decision table shown in the Table 4.9.

Table 4.8 Decision-Making Table of Process Evaluation of Construction Projects

U	Conditions (C)										Decision (D) d
	a_1	a_2	a_3	a_4	a_5	a_6	a_7	a_8	a_9	a_{10}	
n_1	96.33	92.2	86.23	88.23	93.7	95.67	89.45	90.45	84.53	90.27	91.49
n_2	91.27	84.21	94.46	92.56	86.87	87.3	83.2	88.23	90.15	86.53	88.32
n_3	98.15	94.37	100	98.51	97.3	95.79	97.43	96.63	96.78	93.76	96.64
n_4	82.3	80.2	75.16	70.43	68.26	64.6	63.2	70.36	63.12	60.89	69.73
n_5	66.27	67.27	65.66	68.26	71.75	69.63	71.36	72.6	69.67	73.12	69.55
n_6	83.42	82.45	86.24	83.15	72.23	73.33	66.27	76.3	77.12	83.32	78.39
n_7	96.1	93.12	90.13	86.12	82.15	91.13	90.1	86.17	91.11	90.15	89.81
n_8	86.42	83.7	82.76	76.31	78.5	80.7	73.19	74.26	72.34	78.67	79.3
n_9	96.13	98.16	95.76	97.68	86.96	96.61	95.7	86.63	91.7	83.27	93.22

n_{10}	89.22	93.75	96.26	93.34	86.75	97.9	90.23	91.7	84.81	83.78	91.8
n_{11}	82.72	83.46	81.7	86.33	76.36	80.2	78.64	85.71	86.24	89.3	82.97
n_{12}	95.29	92.14	86.76	88.4	81.23	86.53	82.5	80.79	93.87	90.67	87.69
n_{13}	86.3	95.2	91.24	96.23	85.74	93.21	88.93	86.3	89.43	95.65	91.42
n_{14}	96.1	86.23	85.16	90.67	83.27	90.62	88.12	83.46	82.2	82.76	87.38
n_{15}	90.12	81.9	80.86	81.21	72.97	81.79	72.64	72.43	71.34	80.9	79.51
n_{16}	84.32	82.63	80.67	81.56	73.35	83.4	73.26	78.91	76.61	70.61	79.26
n_{17}	90.56	89.65	88.43	90.43	89.17	91.73	94.66	93.67	87.32	87.56	90.48
n_{18}	60.6	79.21	89.45	86.81	84.9	88.55	90.65	87.4	84	78.1	83.28

Table 4.9 Evaluation Decision Table of Grade Classification

U	Condition (C)										Decision (D) d
	a_1	a_2	a_3	a_4	a_5	a_6	a_7	a_8	a_9	a_{10}	
n_1	Excellent	Excellent	Good	Good	Excellent	Excellent	Good	Excellent	Good	Excellent	Excellent
n_2	Excellent	Good	Excellent	Excellent	Good	Good	Good	Good	Excellent	Good	Good
n_3	Excellent	Excellent	Excellent	Excellent	Excellent	Excellent	Excellent	Excellent	Excellent	Excellent	Excellent
n_4	Good	Good	Eligible	Eligible	Bad	Bad	Bad	Eligible	Bad	Bad	Bad
n_5	Bad	Bad	Bad	Bad	Eligible	Bad	Eligible	Eligible	Bad	Eligible	Bad
n_6	Good	Good	Good	Good	Eligible	Eligible	Bad	Eligible	Eligible	Good	Eligible
n_7	Excellent	Excellent	Excellent	Good	Good	Excellent	Excellent	Good	Excellent	Excellent	Good
n_8	Good	Good	Good	Eligible	Eligible	Good	Eligible	Eligible	Eligible	Eligible	Eligible
n_9	Excellent	Excellent	Excellent	Excellent	Good	Excellent	Excellent	Good	Excellent	Good	Excellent

n_{10}	Good	Excellent	Excellent	Excellent	Good	Excellent	Excellent	Excellent	Good	Good	Excellent
n_{11}	Good	Good	Good	Good	Eligible	Good	Eligible	Good	Good	Good	Good
n_{12}	Excellent	Excellent	Good	Good	Good	Good	Good	Good	Excellent	Excellent	Good
n_{13}	Good	Excellent	Excellent	Excellent	Good	Excellent	Good	Good	Good	Excellent	Excellent
n_{14}	Excellent	Good	Good	Excellent	Good	Excellent	Good	Good	Good	Good	Good
n_{15}	Excellent	Good	Good	Good	Eligible	Good	Eligible	Eligible	Eligible	Good	Eligible
n_{16}	Good	Good	Good	Good	Eligible	Good	Eligible	Eligible	Eligible	Eligible	Eligible
n_{17}	Excellent	Good	Good	Excellent	Good	Excellent	Excellent	Excellent	Good	Good	Excellent
n_{18}	Bad	Eligible	Good	Good	Good	Good	Excellent	Good	Good	Eligible	Good

4.3.1.2 Search of Reduct and Establishment of Preferential Rules

As to the condition attributes and decision attributes, obviously, we can see in Table 4.9 that excellent is better than good, good is better than eligible, and eligible is better than bad. These mean excellent > good > eligible > bad. That is to say, the attributes can be regarded as preference attributes. According to the decision attributes, the comprehensive evaluation can be divided into four preferential ordinal classes: Cl_1 = {*bad*}, Cl_2 = {*eligible*}, Cl_3 = {*good*}, and Cl_4 = {*excellent*}. Thus, the unions of decision classes are as follows:

$Cl_1^{\leq} = Cl_1$, comprehensive evaluation is bad.
$Cl_2^{\leq} = Cl_1 \cup Cl_2$, comprehensive evaluation is at most eligible.
$Cl_2^{\geq} = Cl_2 \cup Cl_3 \cup Cl_4$, comprehensive evaluation is at least eligible.
$Cl_3^{\leq} = Cl_1 \cup Cl_2 \cup Cl_3$, comprehensive evaluation is at most good.
$Cl_3^{\geq} = Cl_3 \cup Cl_4$, comprehensive evaluation is at least good.
$Cl_4^{\geq} = Cl_4$, comprehensive evaluation is excellent.

We can find 22 reducts based on genetic algorithms, and a satisfactory reduct with minimal number of attributes to be found is $\{a_2, a_4, a_6, a_8\}$, by which the minimal preferential decision rule set $D_{\geq}$ and $D_{\leq}$ extracted are, respectively, shown in Tables 4.10 and 4.11.

Table 4.10 $D_{\geq}$ Probabilistic Decision Rule Table

Rules	*Support*
If $a_2 \geq$ excellent and $a_4 \geq$ good and $a_6 \geq$ excellent and $a_8 \geq$ excellent, then d = excellent	3
If $a_2 \geq$ good and $a_4 \geq$ excellent and $a_6 \geq$ excellent and $a_8 \geq$ excellent, then d = excellent	2
If $a_2 \geq$ excellent and $a_4 \geq$ excellent and $a_6 \geq$ excellent and $a_8 \geq$ good, then d = excellent	1
If $a_2 \geq$ eligible and $a_4 \geq$ good and $a_6 \geq$ good and $a_8 \geq$ good, then $d \geq$ good	6
If $a_2 \geq$ good and $a_4 \geq$ good and $a_6 \geq$ eligible and $a_8 \geq$ eligible, then $d \geq$ eligible	3
If $a_2 \geq$ good and $a_4 \geq$ eligible and $a_6 \geq$ eligible and $a_8 \geq$ eligible, then $d \geq$ eligible	1
If $a_2 \leq$ good and $a_4 \leq$ eligible and $a_6 \leq$ bad and $a_8 \leq$ eligible, then d = bad	2

Table 4.11 $D_{\leq}$ Probabilistic Decision Rule Table ($\beta = 0.95$)

Rules	*Support*
If $a_2 \geq$ excellent and $a_4 \geq$ good and $a_6 \geq$ excellent and $a_8 \geq$ excellent, then $d =$ excellent	3
If $a_2 \geq$ excellent and $a_4 \geq$ excellent and $a_6 \geq$ excellent and $a_8 \geq$ good, then $d =$ excellent	1
If $a_2 \geq$ good and $a_4 \geq$ excellent and $a_6 \geq$ excellent and $a_8 \geq$ excellent, then $d =$ excellent	2
If $a_2 \leq$ excellent and $a_4 \leq$ good and $a_6 \leq$ excellent and $a_8 \leq$ good, then $d \leq$ good	4
If $a_2 \leq$ good and $a_4 \leq$ excellent and $a_6 \leq$ excellent and $a_8 \leq$ good, then $d \leq$ good	2
If $a_2 \leq$ good and $a_4 \leq$ good and $a_6 \leq$ good and $a_8 \leq$ eligible, then $d \leq$ eligible	4
If $a_2 \leq$ good and $a_4 \leq$ eligible and $a_6 \leq$ bad and $a_8 \leq$ eligible, then $d =$ bad	2

From Tables 4.10 and 4.11, we can see that 18 post-evaluations of projects can all be correctly classified according to the preferential rule. That is, the classification quality is 100 percent, which is a satisfying result.

For different post-evaluations of construction projects, there are different focuses in the evaluation process. For example, the focus of the foundation–construction projects should be laid on the quality of the construction of projects, social benefits, environmental protection, and so on. The evaluation of risking-investment projects should focus on the prospect and the referring technology of the projects. The general industrial projects should focus on the evaluation of trial production, the quality of construction, and the design estimate of budget. In practice, we can flexibly produce corresponding preferential decision model combining the characteristics of certain projects with professional knowledge.

4.3.2 Performance Evaluation of Discipline Construction in Teaching–Research Universities Based on Dominance-Based Rough Set

Discipline construction is a comprehensive construction that universities combine teaching staff, disciplinary platform, scientific research, fundamental condition, personnel training, and academic exchange into an integral whole. It involves the whole process of accurate positioning, scientific development planning, clear development goal, and compact subject direction setting; the formation of a high-level academic team and solid disciplinary platform in the university; the cultivation of talents with good quality and the spirit of innovation; and the creation of high-level scientific research achievements. Discipline construction performance shows the overall academic and teaching level and comprehensive power of a university, and has become the foundation of the sustainable development of a university. Performance evaluation of discipline construction is a systematic project. A scientific and reasonable performance evaluation possesses such functions as orientation, identification, improvement, motivation, and management. Although different universities have different demands on discipline construction, the contents of their discipline construction should be basically identical and the evaluation index system of the performance evaluation of discipline construction should be an organic whole. Because the elements affecting the discipline construction performance are complicated, there is no generally acknowledged evaluation system and approach to do the performance evaluation of discipline construction at present. We applied dominance-based rough set into the performance evaluation of discipline construction of a key university, and the rules obtained can finally provide instructive advices.

4.3.2.1 The Basic Principles of the Construction of Evaluation Index System

Whether the evaluation index system is scientific and reasonable or not will directly affect scientificalness and rationality of the results of evaluation, and further affect the realization of research goals. According to the general rules of the performance evaluation of the discipline construction evaluation and goals of the "Tenth Five-year Plan" discipline construction of a university, the performance evaluation index system should obey the following principles:

1. *Completeness*: The index system of performance evaluation of discipline construction should cover every aspect of discipline construction as possible as it can. All indexes selected should be representative, connected, supplementary, and clear to make the system an organic whole.
2. *Scientificalness*: The scientificalness is an important symbol of reliability. The index system of performance evaluation of discipline construction projects should be based on a certain scientific theory and relative documents and research results available domestically and abroad. Ideas from experts in and outside the college should be collected at the same time. What's more, the indexes selected should be able to effectively reflect the achievements and problems in the construction with the least artificial interference to ensure the scientificalness and authority of the index.
3. *Systematization*: Because there are broad elements that affect the discipline construction achievements, we should adopt the principles such as system design and system evaluation and set reasonable structure of indexes to

comprehensively and systematically reflect the achievements and to provide necessary data for evaluation.

4. *Independence*: The index system of performance evaluation of discipline construction projects should be composed of a group of indexes tightly connected, and each index should have certain connotation and extension, and strong pertinence. Meanwhile each index possesses relative independence in the index system.
5. *Feasibility*: The feasibility of the index system means it should be rather operable. Some indexes are very suitable; however, the basic data is unreachable, thus lack of feasibility. If these indexes are selected, they will cause big trouble for evaluation.
6. *Comparability*: The achievements of the subject construction are, to some extent, based on the comparative standards selected. In order to make the system truly and objectively reflect the objective level we reach in the "Tenth Five-Year Plan" discipline construction of our university, we should keep the index system being of comparability both transverse and longitudinal.

4.3.2.2 The Establishment of Index System and Determination of Weight and Equivalent

On the basis of studying and analyzing related materials home and abroad, referring to the existing index system of performance evaluation of the discipline construction in teaching and researching universities and research universities, our team has preliminarily built the index system of performance evaluation of discipline construction considering the actual situation of key universities in our country. Then, we adjust and perfect the system which now contains 6 first-level indices and 57 second-level indices based on many experts and learners' advices through three times Delphi inquiring approaches. Our team broke down the system into questionnaires of the weights of first-level indices and equivalents of second-level indexes and did a survey. The participants were academic leaders, professors, associate professors, teachers, teaching administrators, excellent managers from governmental management departments of scientific research, and so on. We totally handed out 260 questionnaires and received 251, while we got effective 246 questionnaires after subtracting incomplete and ineffective questionnaires, and the rate of effective questionnaires is 94.6 percent.

After analyzing these questionnaires, we can know that each index or weight obeys normal distribution, because the number of research indexes is large, so that we just give the frequency distribution characteristic charts of the first-level index weight questionnaires, and they are shown in Figures 4.1 through 4.6, while each strip height represents corresponding frequency of this group in the figures.

Through making statistical analysis of these questionnaires, we can determine weight of each first-level index and equivalent of each second-level index, and finally construct the index system of performance evaluation of discipline construction shown in Table 4.12.

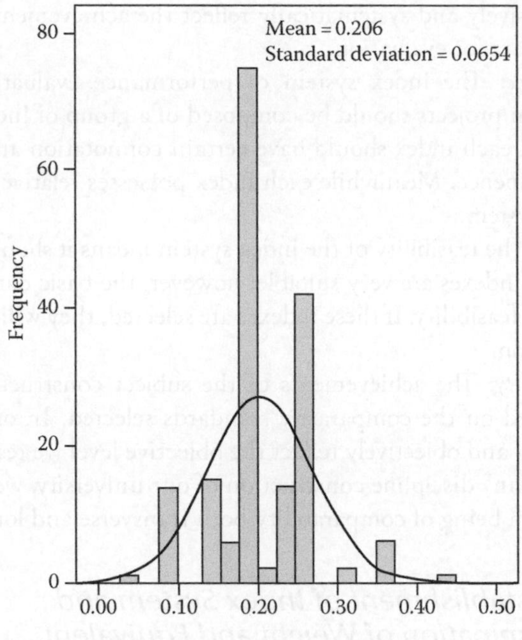

Figure 4.1 Frequency distribution of teaching staff.

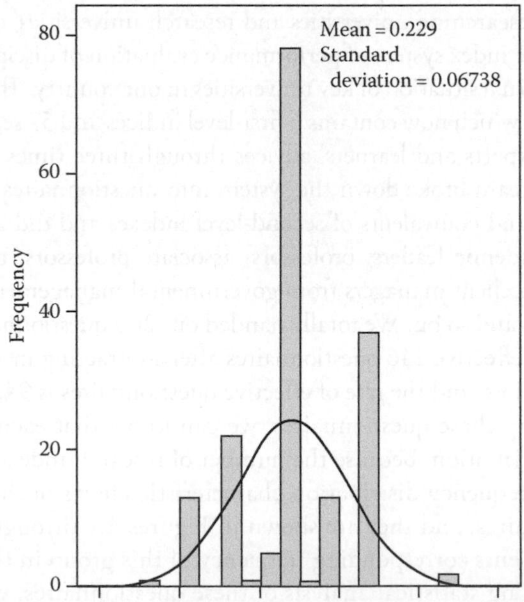

Figure 4.2 Frequency distribution of scientific research.

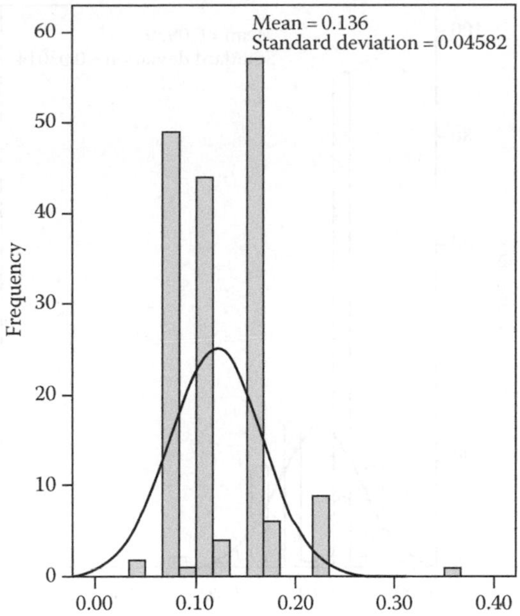

Figure 4.3 Frequency distribution of personnel training.

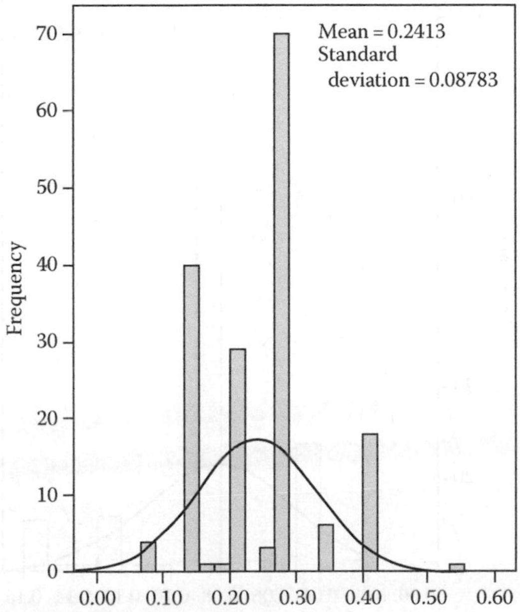

Figure 4.4 Frequency distribution of disciplinary platform.

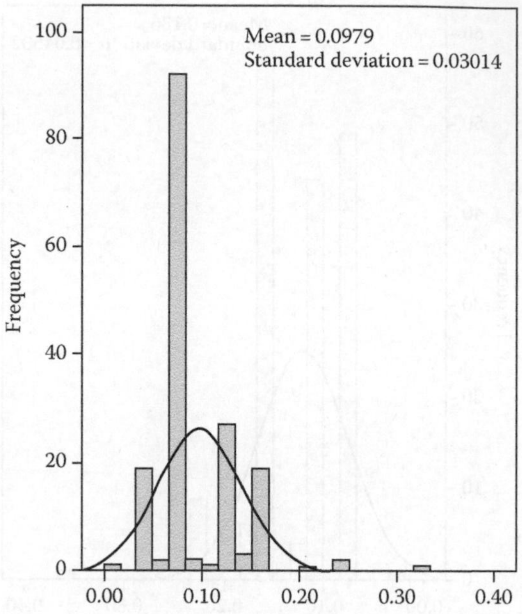

Figure 4.5 Frequency distribution of conditional construction.

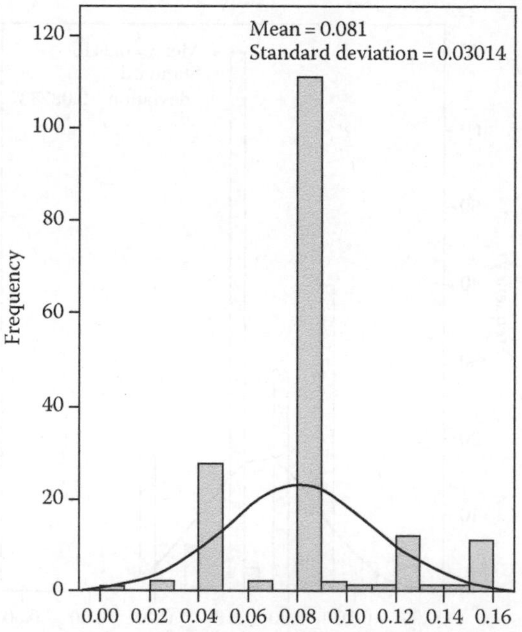

Figure 4.6 Frequency distribution of academic exchange.

Table 4.12 Index System of Performance Evaluation of Discipline Construction and Corresponding Weight and Equivalents

First-Level Indices	*Weights*	*Second-Level Indices*	*Equivalents*	*First-Level Indices*	*Weights*	*Second-Level Indices*	*Equivalents*
Teaching staff	0.2	Academician	7	Academic exchange	0.06	Host an international academic conference	2
		Yangtze river scholars	3.5			Host a national academic conference	1
		Talents training of national level	3			Participate in international academic conference (one person time)	0.1
		PhD supervisor	1.5			Act as national academic chief and vice president (one person time)	1.2

(continued)

Table 4.12 (continued) Index System of Performance Evaluation of Discipline Construction and Corresponding Weight and Equivalents

First-Level Indices	*Weights*	*Second-Level Indices*	*Equivalents*	*First-Level Indices*	*Weights*	*Second-Level Indices*	*Equivalents*
		Professor	1			Act as national or province-level director of academic society (one person time)	0.4
		Talents training of province level	1.2				
		Teachers of doctoral degree	0.7				
		Other teachers	0.5				
Disciplinary platform	0.3	State key subject	3	Talents training	0.12	Per recruiting an undergraduate	0.5
		Key subjects at province level	1.5			Per recruiting a postgraduate (containing engineering postgraduates)	1
		State key labs (including base and center)	2			Per recruiting a doctor	2

		Key labs at province level (including base and center)	1.2			Per graduate	0.5
		First-level subjects for doctors	1			Per postgraduate (containing engineering postgraduates)	1.2
		Second-level subjects for doctors	0.5			Per doctor graduate	3
		National innovative groups	1.5			Per getting a prize of national 100 excellent doctoral dissertation	10
		Innovative groups at province level	1			Per getting a provincial prize of excellent thesis of postgraduate or doctor	5
		School-level innovative groups	0.6			Per getting a national reward	4
		Postgraduates	0.1			Per getting a provincial reward	2

(*continued*)

Table 4.12 (continued) Index System of Performance Evaluation of Discipline Construction and Corresponding Weight and Equivalents

First-Level Indices	*Weights*	*Second-Level Indices*	*Equivalents*	*First-Level Indices*	*Weights*	*Second-Level Indices*	*Equivalents*
						Per getting a national innovative fund	4
						Per getting a provincial innovative fund	2
Scientific research	0.24	SCI or SSCI	1.5	Construction of condition	0.08	Per forming 100,000 fixed funds	1
		EI	1				
		ISTP	0.7				
		Key core	1				
		Normal core	0.7				
		Retrieval periodicals	0.3				
		Non-retrieval periodicals	0.06				
		Compositions	3				

		Teaching materials of main editing	1				
		Reference books of main editing	0.5				
		National scientific research achievements	15				
		Scientific research achievements of province level	6				
		National projects	5				
		Projects of province level	3				
		Transverse research funds of arts (per research is 100,000 yuan)	4.8				
		Vertical research funds of arts (per research is 100,000 yuan)	4.8				

(continued)

Table 4.12 (continued) Index System of Performance Evaluation of Discipline Construction and Corresponding Weight and Equivalents

First-Level Indices	*Weights*	*Second-Level Indices*	*Equivalents*	*First-Level Indices*	*Weights*	*Second-Level Indices*	*Equivalents*
		Transverse research funds of science (per research is 100,000 yuan)	1.2				
		Vertical research funds of science (per research is 100,000 yuan)	1.2				
		Patent licensing	2				
		National teaching achievements	6				
		Teaching achievements at province level	3				

4.3.2.3 Data Collection and Pretreatment

According to the index system of evaluation, we collected a key university's discipline construction materials of 11 collages from January 2001 to December 2005 and calculated the scores of first- and second-level indexes of each collage. We only dealt with and analyzed the data of first-level indexes, and second-level indices were done similarly. To make these indexes possess comparability, we carried out normalization processing of these materials according to the scores of the indexes, while normalized formula is as follows:

$$a'_{ij} = \frac{a_{ij}}{a^{\bullet}_j} \times 100$$

where

a_{ij} represents marks of first-level indexes number j in i collage
$a^{\bullet}_j$ represents the maximum marks of first-level indexes number j in all collages, $i = 1, \ldots, 11, j = 1, \ldots, 6$

We can obtain the benefits of each first-level index of each collage with the products of each index value normalized and corresponding weight of first-level indexes, and then get the comprehensive benefits by summing the accumulation of every index benefits of each collage. The decision table constructed of performance evaluation of discipline construction is shown in Table 4.13.

4.3.2.4 Data Discretization

The equal frequency discretization approach is used to discrete condition attribute values and decision attribute values of the 11 collages, which are divided into three different preference intervals: excellent, good, and bad, as shown in Table 4.14.

According to discretization grade of each attribute in Table 4.14, the data in it can be discretized to a preferential decision table shown in Table 4.15.

4.3.2.5 Search of Reducts and Generation of Preferential Rules

For those conditions and decision attributes in Table 4.15, obviously, excellent is better than good, while good is better than bad. That means excellent > good > bad. That is to say, these attributes exist with preference information. According to decision attributes, the comprehensive performance evaluation of discipline construction can be divided into three preferential ordinal classes: $Cl_1 = \{bad\}$, $Cl_2 = \{good\}$, $Cl_3 = \{excellent\}$, and after dividing the universe according to preferential decision class, we can obtain the following union of decision classes: for

Table 4.13 Decision Table of Performance Evaluation of Discipline Construction

Units	*Teaching Staff*	*Disciplinary Platform*	*Research*	*Talents Training*	*Condition*	*Academic Exchange*	*Comprehensive Benefits*
A collage	21	24	23	11	10	8	97
B collage	6.9	6	8.2	6.9	2.5	1.7	32.1
C collage	12.6	9.7	13.7	14	2.8	3.3	56.1
D collage	9.3	6.7	5	13.6	1	0.8	36.4
E collage	12.9	9	11.2	13.2	17	3.3	51.3
F collage	7.3	1.8	2.2	3.4	2.1	1.1	18
G collage	4.4	2.2	3.1	4	0.8	0.1	14.6
H collage	6.8	0.8	1.4	2.7	0.4	0.9	13
I collage	6.3	2.7	3.5	8.2	0.4	2	23
J collage	9.9	1	2.3	4.7	0.1	0.9	18.9
K collage	2.3	0.3	0.2	0.7	0.1	0.3	3.8

Table 4.14 Preference Discretization Intervals of Performance Evaluation of Discipline Construction

Level	*Teaching Staff*	*Disciplinary Platform*	*Scientific Research*	*Talents Training*	*Condition Construction*	*Academic Exchange*	*Comprehensive Efficiency*
Excellent	[12.4, *)	[7.5, *)	[9.7, *)	[12.1, *)	[2.3, *)	[2.7, *)	[44.3, *)
Good	[7.6, 12.4)	[2.0, 7.5)	[2.7, 9.7)	[4.4, 12.1)	[0.6, 2.3)	[1.0, 2.7)	[19.8, 43.9)
Bad	[*, 7.6)	[*, 2)	[*, 2.7)	[*, 4.4)	[*, 0.6)	[*, 1.0)	[*, 19.8)

Table 4.15 Preferential Table of Performance Evaluation of Discipline Construction

Collages	*Teaching Staff*	*Disciplinary Platform*	*Scientific Research*	*Talents Training*	*Condition Construction*	*Academic Exchange*	*Comprehensive Evaluation*
A collage	Excellent	Excellent	Excellent	Good	Excellent	Excellent	Excellent
B collage	Bad	Good	Good	Good	Excellent	Good	Good
C collage	Excellent	Excellent	Excellent	Excellent	Excellent	Excellent	Excellent
D collage	Good	Good	Good	Excellent	Good	Bad	Good
E collage	Excellent	Excellent	Excellent	Excellent	Good	Excellent	Excellent
F collage	Good	Bad	Bad	Bad	Good	Good	Bad
G collage	Bad	Good	Good	Bad	Good	Bad	Bad
H collage	Good	Bad	Bad	Bad	Bad	Bad	Bad
I collage	Bad	Good	Good	Good	Bad	Good	Good
J collage	Good	Bad	Bad	Good	Bad	Bad	Good
K collage	Bad	Bad	Bad	Bad	Bad	Bad	Bad

$Cl_1^{\leq} = Cl_1$, the comprehensive performance evaluation of discipline construction is bad; for $Cl_2^{\leq} = Cl_1 \cup Cl_2$, the comprehensive performance evaluation of discipline construction is at most good; for $Cl_2^{\geq} = Cl_2 \cup Cl_3$, the comprehensive performance evaluation of discipline construction is good at least; for $Cl_3^{\leq} = Cl_1 \cup Cl_2 \cup Cl_3$, the comprehensive performance evaluation of discipline construction is at most excellent; and for $Cl_3^{\geq} = Cl_3$, the comprehensive performance evaluation of discipline construction is excellent. According to dominance-based rough sets, we can search four reducts: {scientific research, talents training}, {disciplinary platform, talents training}, {talents training, academic exchange}, {teaching staff, talents training}, while the minimum preferential decision rules sets $D_{\geq}$ and $D_{\leq}$ generated from the reducts are shown in Tables 4.16 and 4.17.

4.3.2.6 Analysis of Evaluation Results

We can see from the above preferential rules that classification quality of the four reducts is 100 percent, that is, all objects are correctly classified based on the four generated reductions. For reducts {scientific research, talents training} ∩ {disciplinary platform, talents training} ∩ {talents training, academic exchange} ∩ {teaching

Table 4.16 Preferential Decision Rules Sets $D_{\geq}$ Generated from the Reduct {Scientific Research, Talents Training}

Preferential Decision Rules	*Support Numbers*
Scientific research = excellent AND talents training >= good ⇒ comprehensive efficiency = excellent	3
Scientific research >= bad AND talents training >= good ⇒ comprehensive efficiency >= good	4
Scientific research <= good AND talents training = bad ⇒ comprehensive efficiency = bad	4

Table 4.17 Preferential Decision Rules Sets $D_{\leq}$ Generated from the Reduct {Scientific Research, Talents Training}

Preferential Decision Rules	*Support Numbers*
Scientific research = excellent AND talents training >= good ⇒ comprehensive efficiency = excellent	3
Scientific research <= good AND talents training <= excellent ⇒ comprehensive efficiency <= good	4
Scientific research <= good AND talents training = bad ⇒ comprehensive efficiency = bad	4

staff, talents training} = {talents training}, so talents training is the core. From Tables 4.18 through 4.20, we can see that the comprehensive performance evaluation of discipline construction is in the middle place if the benefit of the talents training is moderate despite of the performance of the indexes. It is obvious that talents training is a central task of the discipline construction.

Table 4.18 Preferential Decision Rules Sets $D_{\geq}$ Generated from the Reduct {Teaching Staff, Talents Training}

Preferential Decision Rules	*Support Numbers*
Teaching staff = excellent AND talents training >= good ⇒ comprehensive training = excellent	3
Teaching staff >= bad AND talents training >= good ⇒ comprehensive efficiency >= good	4
Teaching staff < good AND talents training = bad ⇒ comprehensive efficiency = bad	4

Table 4.19 Preferential Decision Rules Sets $D_{\leq}$ Generated from the Reduct {Teaching Staff, Talents Training}

Preferential Decision Rules	*Support Numbers*
Teaching staff >= excellent AND talents training >= good ⇒ comprehensive efficiency = excellent	3
Teaching staff <= good AND talents training <= excellent ⇒ comprehensive efficiency <= good	4
Teaching staff <= good AND talents training = bad ⇒ comprehensive efficiency = bad	4

Table 4.20 Preferential Decision Rules Sets $D_{\geq}$ Generated from the Reduct {Disciplinary Platform, Talents Training}

Preferential Decision Rules	*Support Numbers*
Disciplinary platform >= excellent AND talents training >= good ⇒ comprehensive efficiency = excellent	3
Disciplinary platform >= bad AND talents training >= good ⇒ comprehensive efficiency >= good	4
Disciplinary platform <= good AND talents training = bad ⇒ comprehensive efficiency = bad	4

Table 4.21 Preferential Decision Rules Sets $D_{\leq}$ Generated from the Reduct {Disciplinary Platform, Talents Training}

Preferential Decision Rules	*Support Numbers*
Disciplinary platform good >= excellent AND talents training >= good ⇒ comprehensive efficiency = excellent	3
Disciplinary platform <= AND talents training <= excellent ⇒ comprehensive efficiency <= good	4
Disciplinary platform <= AND talents training = bad ⇒ comprehensive efficiency = bad	4

Table 4.22 Preferential Decision Rules Sets $D_{\geq}$ Generated from the Reduct {Talents Training, Academic Exchange}

Preferential Decision Rules	*Support Numbers*
Talents training >= good AND academic exchange = excellent ⇒ comprehensive efficiency = excellent	3
Talents training >= AND academic exchange >= bad ⇒ comprehensive efficiency >= good	4
Talents training = bad AND academic exchange <= good ⇒ comprehensive efficiency = bad	4

According to the at least minimum one grade preferential rules sets $D_{\geq}$ and the at most one grade preferential rules sets $D_{\leq}$ generated from the reducts {disciplinary platform, talents training}, {disciplinary platform, talents training}, and {teaching staff, talents training} of Tables 4.20 and 4.21, we can see that if the benefit of scientific research or disciplinary platform is excellent, while the benefit of talents training is good in the discipline construction, the comprehensive performance evaluation of discipline construction can reach excellence; although the benefit of the talents training is excellent, the benefit of disciplinary platform or scientific research is at most good, thus the comprehensive evaluation is at most good. Obviously disciplinary platform, scientific research, and teaching staff play an important role in talents training and the performance of discipline construction.

According to preferential rules sets $D_{\geq}$ and the at most one grade preferential rules sets $D_{\leq}$ generated from the reducts {talents training, academic exchange} of Tables 4.22 and 4.23, we can see that although the benefit of academic exchange is excellent and the benefit of talents training is good, the performance evaluation of discipline construction can still be excellent; although the performance evaluation of talents training is excellent, and academic exchange is bad, the comprehensive performance evaluation of discipline construction can become excellent at most.

Table 4.23 Preferential Decision Rules Sets $D_{\leq}$ Generated from the Reduct {Talents Training, Academic Exchange}

Preferential Decision Rules	*Support Numbers*
Talents training >= good AND academic exchange >= excellent ⇒ comprehensive efficiency (excellent)	3
Talents training <= excellent AND academic exchange = bad ⇒ comprehensive efficiency <= good	2
Talents training <= good AND academic exchange <= good ⇒ comprehensive efficiency <= good	2
Talents training = bad AND academic exchange <= good ⇒ comprehensive efficiency = bad	4

Obviously, academic exchange promotes the performance of the talents training and discipline construction.

From the reducts generated, there exists no condition construction index in the reducts, thus we can conclude that condition construction indexes is redundant attributes. For example, the construction of disciplinary platform may cover condition construction.

The dominance-based RST is used in the evaluation decision of performance of discipline construction, and we can obtain some beneficial rules and knowledge. Then the preferential model generated can not only scientifically classify and evaluate the performance of discipline construction, and reasonably explain the roles that the different indexes play in the discipline construction, but also produce relatively less rules. Such a preferential expression rules can be easier to be understood by different participants in discipline construction.

4.4 Summary

This chapter firstly introduces the dominance-based rough set approach and discusses inconsistence and indiscernibility relation based on dominance relations. Then, it proposes dominance variable precision rough sets that can obtain preferential probability rules from preferential multiple attribute decision tables, meanwhile it explains the approach with an example. Finally, the dominance-based RST is used in the piratical cases of the post-evaluation of construction projects and the performance evaluation of discipline construction.

The rough sets models and the variable precision rough sets models are based on indiscernibility relation, while dominance rough sets models and dominance variable precision rough sets models are based on dominance relations; condition attribute and decision class possess preferential information, and the proximate knowledge is upward union and downward union in the dominance rough sets and dominance variable precision rough sets. Indiscernibility relation is reflexive, symmetric, and transitive, while dominance relation is reflexive and transitive. These are the differences between the approaches of rough sets and variable precision rough sets, dominance rough sets and dominance variable precision rough sets.

Because preferential multiple attribute decision tables require orderable or preferential data, we should first divide the preference attribute into proximate knowledge granularity based on dominance relations, then deduce preferential model which is a certain preferential rules set acting as "if… then…" with examples provided by the decision-maker. These rules can deal with the incompatibility that possibly arises in the preferential multiple attribute decision system as a result of natural grammar of the rules, the established preferential model approaches to natural reasoning of decision problems. Such a preferential rule is easier for users to understand, and help to generate an interactive structure of probability decision preferential model. The results show that we get the minimum probability rules sets under the relatively few amounts of conditions based on dominance-based rough sets, while the rules induced are relatively fewer but stronger. In addition, rules sets proximately induced by the approximation based on the definition of dominance relation give a more systematic knowledge representation contained in decision tables.

Chapter 5

Hybrid of Rough Set Theory and Fuzzy Set Theory

Rough set theory and fuzzy set theory are extensions of classical set theory, and they are related but distinct and complementary theories. Rough set theory is mainly focused on crisp information granulation and its basic concept is indiscernibility (e.g., an indiscernibility between different objects deduced by different attribute values of described objects in the information system), whereas fuzzy set theory is regarded as a mathematical tool for imitating the fuzziness in the human classification mechanism, which mainly deals with fuzzy information granulation. Because of its simplicity and similarity with the human mind, its concept is always used to express quantity data expressed by language and membership functions in the intelligent system. In fuzzy sets, the attribution of elements may be between yes and no. Let us take the example of a beautiful scenery; we cannot simply classify the beautiful scenery into a category between yes and no. For the set of beautiful scenery, there does not exist good and definite border. Fuzzy sets cannot be described with any precise mathematical formula, but it is included in the physical and psychological process of a human's way of thinking, because the physiology of human reasoning never uses any precise mathematical formula during the physical process of reasoning, and fuzzy sets is important in pattern classification. Essentially, both these theories study the problem of information granularity. The rough set theory studies rough non-overlapping type and the rough concept, while the fuzzy set theory studies the fuzziness between overlapping sets, and these naturally lead to investigating the possibility of the "hybrid" between the rough set and the fuzzy set.

5.1 The Basic Concepts of the Fuzzy Set Theory

The notion of fuzzy sets provides a normative tool for acquisition, representation, and processing of fuzziness, which can deal with various problems with vague boundary definition. The fuzzy set theory deals with the ill definition of the boundary of class through a continuous generalization of a set characteristic functions, and the indiscernibility between different objects is not used in the fuzzy set theory. Fuzzy set may be regarded as a class with vague boundaries, whereas rough set is a crisp set described roughly.

5.1.1 Fuzzy Set and Fuzzy Membership Function

In the classical set, let $X \subseteq U$ and $x \in U$, then x may be an element in the given class $X \subseteq U$, or may not be in the given class X, as shown in Figure 5.1a. This property can be defined as the characteristic function μ_X: $U \rightarrow \{0, 1\}$, that is,

$$\mu_X(x) = \begin{cases} 1 & \text{if } x \in X \\ 0 & \text{if } x \notin X \end{cases} \tag{5.1}$$

However, in practical application, the boundary between classes may be overlapping; therefore, it cannot determine whether an input pattern x totally belongs to class X. The concept of fuzzy sets is introduced, as shown in Figure 5.1b, to deal with uncertainty. In the fuzzy set, the concept of characteristic function is revised as fuzzy membership function

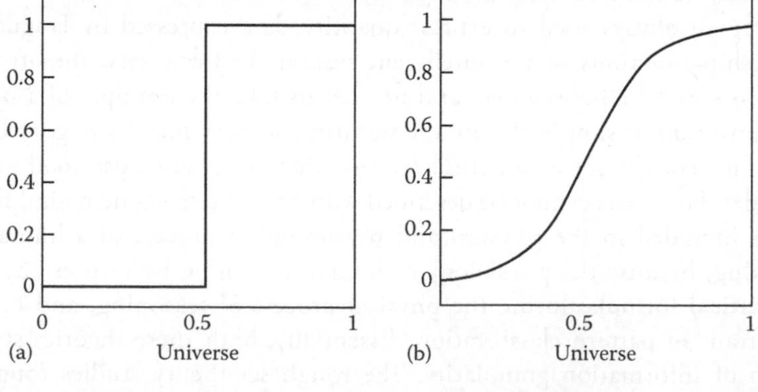

Figure 5.1 Membership functions of the (a) classical set theory and (b) fuzzy set theory.

$$\mu_{FX}: U \to [0,1] \tag{5.2}$$

The larger value in the function indicates larger membership degree that an element belongs to a given set. A fuzzy set in the universe U is fully described by its membership function.

If universe U is a limited or countable set, fuzzy set FX can be expressed as

$$FX = \sum_{i=1}^{n} \frac{\mu_{FX_i}(x_i)}{x_i} \tag{5.3}$$

where

x_i is an element of the fuzzy set FX ($i = 1, 2, \ldots, n$)

μ_{FX_i} is the membership degree of the number i element in the fuzzy set FX

If U is an uncountable and infinite set, then the fuzzy set FX can be expressed as

$$FX = \int_{x \in U} \frac{\mu_{FX}(x)}{x} \tag{5.4}$$

Fuzzy membership functions are mostly subjective, and different people can assign different membership functions to the same value. By allowing the part membership, the fuzzy set theory has extended the classical set, while the introduction of part membership has provided a more realistic model for simulating the ill definition of boundaries of classes. However, it has brought about many essential problems, such as the semantic interpretation of the membership value and the operation of set theory.

Example 5.1 Suppose Age is a universe, U = [0, 100], and Chad has listed the membership functions of the two "old" FO and "young" FY fuzzy subsets as follows:

$$\mu_{FO}(x) = \begin{cases} 0 & 0 \le x \le 50 \\ \left(1 + \left(\dfrac{x-50}{5}\right)^{-2}\right)^{-1} & 50 < x \le 100 \end{cases}$$

$$\mu_{FY}(x) = \begin{cases} 1 & 0 \le x \le 25 \\ \left(1 + \left(\dfrac{x-25}{5}\right)^{2}\right)^{-1} & 25 < x \le 100 \end{cases}$$

Then the membership functions of these two fuzzy subsets are shown in Figure 5.2a and b, respectively.

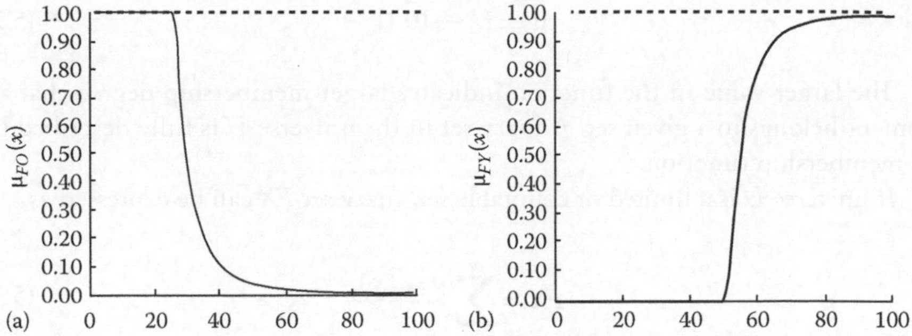

Figure 5.2 The membership functions of (a) the "old" and (b) the "young".

5.1.2 Operation of Fuzzy Subsets

Definition 5.1 Suppose $FX \subseteq U$ and $FY \subseteq U$ are two fuzzy sets in universe. For any $x \in U$, if $\mu_{FX}(x) \leq \mu_{FY}(x)$, then FX is included in FY or FY include FX, and is denoted by $FX \subseteq FY$. If $FX \subseteq FY$ and $FY \subseteq FX$ are both tenable, then FX equals FY, and is denoted by $FX = FY$. Empty set ϕ indicates a fuzzy set whose membership function is constant 0, while universe U indicates a fuzzy set whose membership function is constant 1.

Suppose $FX \cup FY$ is the union of the fuzzy sets FX and FY, then its membership function is defined as

$$\mu_{FX \cup FY}(x) = \mu_{FX}(x) \vee \mu_{FY}(x) = \max\{\mu_{FX}(x), \mu_{FY}(x)\}$$

Suppose $FX \cap FY$ is the intersection of the fuzzy sets FX and FY, then its membership function is defined as

$$\mu_{FX \cap FY}(x) = \mu_{FX}(x) \wedge \mu_{FY}(x) = \min\{\mu_{FX}(x), \mu_{FY}(x)\}$$

Suppose $\neg FX$ or FX^C is the complementary set of the fuzzy set FX, then its membership function is defined as

$$\mu_{\neg FX}(x) = 1 - \mu_{FX}(x)$$

Suppose $FX - FY$ is the difference set of the fuzzy set FX and FY, then its membership function is defined as

$$\mu_{FX-FY}(x) = \mu_{FX \cap \neg FY}(x) = \mu_{FX}(x) \wedge (1 - \mu_{FY}(x))$$

Let $FX \subseteq U$, $FY \subseteq U$, $FZ \subseteq U$ be three fuzzy sets in U, then the operation of these fuzzy sets meets the following properties:

1. Commutative law

$$\begin{cases} FX \cup FY = FY \cup FX \\ FX \cap FY = FY \cap FX \end{cases}$$

2. Distributive law

$$\begin{cases} FX \cup (FY \cap FZ) = (FX \cup FY) \cap (FX \cup FZ) \\ FX \cap (FY \cup FZ) = (FX \cap FY) \cup (FX \cap FZ) \end{cases}$$

3. Associative law

$$\begin{cases} (FX \cup FY) \cup FZ = FX \cup (FY \cup FZ) \\ (FX \cap FY) \cap FZ = FX \cap (FY \cap FZ) \end{cases}$$

4. Absorptive law

$$\begin{cases} FX \cup (FY \cap FX) = FX \\ FX \cap (FY \cup FX) = FX \end{cases}$$

5. Identity law

$$\begin{cases} FX \cup \Phi = FX \\ FX \cap U = FX \end{cases}$$

6. Primitive law

$$\begin{cases} FX \cup U = U \\ FX \cap \Phi = \Phi \end{cases}$$

7. Complementary law

$$\begin{cases} FX \cup (\neg FX) = U \\ FX \cap (\neg FX) = \Phi \end{cases}$$

8. De Morgan law

$$\begin{cases} \neg(FX \cup FY) = (\neg FX) \cap (\neg FY) \\ \neg(FX \cap FY) = (\neg FX) \cup (\neg FY) \end{cases}$$

9. Double negative law

$$\neg(\neg FX) = FX$$

10. Law of idempotence

$$\begin{cases} FX \cup FX = FX \\ FX \cap FX = FX \end{cases}$$

11. Others

$$\begin{cases} \neg \Phi = U \\ \neg U = \Phi \end{cases}$$

The operation between two fuzzy subsets in fact is to do the corresponding operation on the membership degree point by point, and the rules are as follows:

$$FX = \Phi \Leftrightarrow \mathit{for}\ \forall x \in U,\quad \mu_{FX}(x) = 0$$

$$FX = FY \Leftrightarrow \mathit{for}\ \forall x \in U,\quad \mu_{FX}(x) = \mu_{FY}(x)$$

$$\neg FX \Leftrightarrow \mathit{for}\ \forall x \in U,\quad \mu_{\neg FX}(x) = 1 - \mu_{FX}(x)$$

$$FX \subseteq FY \Leftrightarrow \mathit{for}\ \forall x \in U,\quad \mu_{FX}(x) \le \mu_{FY}(x)$$

$$FZ = FX \cup FY \Leftrightarrow \mathit{for}\ \forall x \in U,\quad \mu_{FZ}(x) = \max(\mu_{FX}(x), \mu_{FY}(x))$$

$$FZ = FX \cap FY \Leftrightarrow \mathit{for}\ \forall x \in U,\quad \mu_{FZ}(x) = \min(\mu_{FX}(x), \mu_{FY}(x))$$

If the universe U is an infinite uncountable set, and FX and FY are two random fuzzy sets of U, we can obtain

$$FX \cup FY = \int_{x \in U} \frac{(\mu_{FX}(x) \vee \mu_{FY}(x))}{x}$$

$$FX \cap FY = \int_{x \in U} \frac{(\mu_{FX}(x) \wedge \mu_{FY}(x))}{x}$$

$$\neg FX = \int_{x \in U} \frac{(1 - \mu_{FX}(x))}{x}$$

As for the operation of fuzzy sets, such as union and intersection, it can be extended to any fuzzy set, and the supremum (sup) is often used to replace max or "∨," while the infimum (inf) is often used to replace min or "∧."

5.1.3 Fuzzy Relation and Operation

Given two sets X, Y, a common relation R from X to Y refers to the subset of $X \times Y$. If $(x, y) \in R$, then x and y exists relation R, namely, xRy. A common relation can only describe whether there exists some relation between two elements. But in the real world, there are large and more complex relations, while relations between elements are not simply in presence or absence but in different degrees, so the concept of fuzzy relation is introduced to describe the complex relation.

Definition 5.2 Name a fuzzy subset R of Cartesian product $X \times Y$ a fuzzy relation from X to Y. If $(x, y) \in X \times Y$, $\mu_R(x, y)$ is regarded as the degree of relation R between u and v, $\mu_R(x, y)$ can be abbreviated as $R(x, y)$. When $U = V$, R is a fuzzy relation on U.

When take only $\mu_R(x, y)$ as 0 or 1, R will degrade into a common relation; therefore, the fuzzy relation is the extension of the common relation. The operation of fuzzy relation and fuzzy set is same.

Suppose R_1, R_2 are fuzzy relation on $X \times Y$, then the fuzzy relation meets the following properties:

1. $R_1 \cup R_2 \Leftrightarrow$ for any $(x, y) \in X \times Y$, $\mu_{R_1 \cup R_2}(x, y) = \max(\mu_{R_1}(x, y), \mu_{R_2}(x, y)$.
2. $R_1 \cap R_2 \Leftrightarrow$ for any $(x, y) \in X \times Y$, $\mu_{R_1 \cap R_2}(x, y) = \min(\mu_{R_1}(x, y), \mu_{R_2}(x, y))$.
3. $R_1 \subseteq R_2 \Leftrightarrow$ for any $(x, y) \in X \times Y$, $\mu_{R_1}(x, y) \leq \mu_{R_2}(x, y)$.
4. $\neg R_1 \Leftrightarrow$ for any $(x, y) \in X \times Y$, $\mu_{\neg R_1}(x, y) = 1 - \mu_{R_1}(x, y)$.
5. If the fuzzy relation R on $X \times Y$ is given, for $\forall x, y \in X \times Y$, $\mu_{R^T}(x, y) = \mu_R(x, y)$, then R^T is called as the inverse relation of R.
6. If the given fuzzy relation R on X meets $R \Leftrightarrow \mu_R(x, y) = \begin{cases} 1 & \text{if } x = y \\ 0 & \text{if } x \neq y \end{cases}$, then R is called as the "identical relation" on X.

7. If the given fuzzy relation R on $X \times Y$ meets $R \Leftrightarrow \mu_R(x, y) = 0$, for $\forall(x, y) \in X \times Y$, then R is the "null relation" on $X \times Y$.
8. If the given fuzzy relation R on $X \times Y$ meets, $R \Leftrightarrow \mu_R(x, y) = 1$, for $\forall(x, y) \in X \times Y$, then R is called as the "universal relation" on $X \times Y$.

5.1.4 Synthesis of Fuzzy Relations

Synthesis of fuzzy relations is a kind of extension that is a synthesis of common relations. There are many examples about the synthesis of common relations, such as "grandparent and grandchild" is the synthesis of "father and son" between "father and son."

Definition 5.3 Suppose the given sets X, Y, Z, I_1 is the common relation on $X \times Y$, and I_2 is the common relation on $Y \times Z$, then the synthesis relation I_3 of I_1 and I_2 is $I_3 = I_1 \circ I_2$, which is just a relation from X(by Y) to Z.

Promoting such synthetic method to the fuzzy relation, we can get Definition 5.4.

Definition 5.4 Suppose the given sets X, Y, Z, I_1 is the common relation on $X \times Y$, and I_2 is the common relation on $Y \times Z$, then the synthesis relation from R_1 to R_2 is a fuzzy relation from X to Z, denoted by $R_1 \circ R_2$, and its membership function is $\mu_{R_1 \circ R_2}(x,z) = \bigvee_{y \in Y} (\mu_{R_1}(x,y) \wedge \mu_{R_2}(y,z))$.

In the fuzzy relation R, each pair of the elements (x, y) is corresponding to a value $R(x, y)$ that ranges between 0 and 1, and represents that x has the degree of relation R to y, or it is called the corresponding degree of x to y for the relation R.

Example 5.2 Suppose $R_1 = \begin{bmatrix} 0.4 & 0.6 & 0.3 \\ 1 & 0 & 0.4 \\ 0 & 0.5 & 1 \\ 0.6 & 0.7 & 0.8 \end{bmatrix}$, $R_2 = \begin{bmatrix} 0.1 & 0.9 \\ 0.9 & 0.1 \\ 0.6 & 0.4 \end{bmatrix}$, please solve the synthesis R_3 of R_1 to R_2.

Solution: $R_3 = R_1 \circ R_2$, according to Definition 5.3, the first row and the first line element r_{11} in R_3 is

$$\begin{aligned} r_{11} &= (0.4 \wedge 0.1) \vee (0.6 \wedge 0.9) \vee (0.3 \wedge 0.6) \\ &= 0.1 \vee 0.6 \vee 0.3 \\ &= 0.6 \end{aligned}$$

Similarly, we can get other elements in R_3, then R_3 is

$$R_3 = R_1 \circ R_2 = \begin{bmatrix} 0.6 & 0.4 \\ 0.4 & 0.9 \\ 0.6 & 0.4 \\ 0.7 & 0.6 \end{bmatrix}$$

5.1.5 λ-Cut Set and the Decomposition Proposition

In practical application, it is often necessary to make precise judges for fuzziness, while λ-cut set is a very important concept in the reciprocal conversion between the fuzzy set and the ordinary set, and frequently used in the fuzzy decision.

Definition 5.5 Suppose fuzzy set $FX \subseteq U$, for $\forall\lambda \in [0, 1]$, common set $FX_\lambda = \{x \in U \mid \mu_{FX}(x) \geq \lambda\}$ is the λ-cut set of FX, while λ is called level. $FX_\lambda = \{x \in U \mid \mu_{FX}(x) > \lambda\}$ is regarded as the strong λ-cut set of FX.

According to Definition 5.5, we know that, for every λ-cut set, a common subset FX_λ can be determined. Obviously, we can convert fuzzy set into common set through λ-cut set.

Example 5.3 Suppose $U = \{3, 2, 5, 6, 9, 10)$, the fuzzy set FX in U is

$$FX = \frac{0.2}{3} + \frac{0.4}{2} + \frac{0.6}{5} + \frac{0.9}{9} + \frac{1}{10},$$

then $FX_{0.2} = \{3, 2, 5, 9, 10\}$, $FX_{0.6} = \{5, 9, 10\}$, $FX_{0.9} = \{9, 10\}$, $FX_1 = \{10\}$.

Suppose the fuzzy set $FX \subseteq U$, $FY \subseteq U$, while the λ-cut set of the fuzzy set has the following properties:

1. $(FX \cup FY)_\lambda = FX_\lambda \cup FY_\lambda$
2. $(FX \cap FY)_\lambda = FX_\lambda \cap FY_\lambda$
3. If $\lambda \in [0, 1]$, $\mu \in [0, 1]$, and $\lambda < \mu$, then $FX_\lambda \supseteq FX_\mu$

According to property (3), we could get the result that the lower level of the cut set is, the larger FX_λ is; the higher level of the cut set is, the smaller FX_λ is. Obviously, when $\lambda = 1$, $FX_\lambda = 1$ is minimal; if $FX_{\lambda=1} \neq \Phi$, then it is the core of FX. The following definitions hereby are given.

Definition 5.6 Suppose $FX \subseteq U$, then

1. $Cor(FX) = \{x \in U \mid \mu_{FX}(x) = 1\}$ is called the core of FX.
2. $Supp(FX)_\lambda = \{x \in U \mid \mu_{FX}(x) > 0\}$ is called the support of FX.

3. If $Cor(FX) \neq \Phi$, then FX is the regular fuzzy set, otherwise the irregular fuzzy set.
4. $Hgt(FX) = \sup_{x \in U} \mu_{FX}(x)$ is called the height of FX.

Definition 5.7 Suppose the fuzzy set $FX \subseteq U$, $\lambda \in [0, 1]$, while λFX_λ is the product of the value λ and FX_λ, and it is a fuzzy set and its membership function is as follows:

$$\mu_{\lambda FX_\pi}(x) = \begin{cases} \lambda (x \in FX_\lambda) \\ 0 (x \notin FX_\lambda) \end{cases}$$

Proposition 5.1 Suppose the fuzzy set $FX \subseteq U$, then FX can be decomposed into the union of a series of the scalar product fuzzy set λFX_λ, that is, $FX = \bigcup_{\lambda \in [0,1]} \lambda A_\lambda$.

Example 5.4 Suppose $U = \{x_1, x_2, x_3, x_4, x_5\}$, $FX \subseteq U$, while each cut set of FX is as follows:

$$FX_\lambda = \begin{cases} \{x_1, x_2, x_3, x_4, x_5\} & 0 \le \lambda \le 0.3 \\ \{x_1, x_2, x_3, x_5\} & 0.3 < \lambda \le 0.6 \\ \{x_1, x_3, x_5\} & 0.6 < \lambda \le 0.7, \\ \{x_1, x_3\} & 0.7 < \lambda \le 0.8 \\ \{x_3\} & 0.8 < \lambda \le 1 \end{cases}$$ and calculate the fuzzy set FX.

Solution: According to Proposition 5.1, we can get

$$FX = \bigcup_{\lambda \in [0,1]} \lambda A_\lambda = 0.3A_{0.3} \cup 0.6A_{0.6} \cup 0.7A_{0.7} \cup 0.8A_{0.8} \cup 1A_1$$

$$= \left(\frac{0.3}{x_1} + \frac{0.3}{x_2} + \frac{0.3}{x_3} + \frac{0.3}{x_4} + \frac{0.3}{x_5}\right) \cup \left(\frac{0.6}{x_1} + \frac{0.6}{x_2} + \frac{0.6}{x_3} + \frac{0.6}{x_5}\right) \cup \left(\frac{0.7}{x_1} + \frac{0.7}{x_3} + \frac{0.7}{x_5}\right)$$

$$\cup \left(\frac{0.8}{x_1} + \frac{0.8}{x_3}\right) \cup \frac{1}{x_3}$$

$$= \frac{0.8}{x_1} + \frac{0.6}{x_2} + \frac{1}{x_3} + \frac{0.3}{x_4} + \frac{0.7}{x_5}$$

5.1.6 The Fuzziness of Fuzzy Sets and Measure of Fuzziness

Although we can get a comprehensive understanding of a fuzzy set by fuzzy membership function, it still needs some quantitative indexes to describe the fuzzy set if we want to characterize features of the fuzzy set in a certain aspect, or compare different fuzzy sets.

Definition 5.8 Suppose the fuzzy set $FX \subseteq U$, $x_i \in U$, $i = 1, 2, \ldots, n$, then the base of FX is

$$|FX| = \sum_{i=1}^{n} \mu_{FX}(x_i)$$

According to Definition 5.8, we know that $|FX|$ is a quantitative index that depicts the "capacity" of FX. When FX is a common set, $|FX|$ becomes the base of the common set.

In 1972, Delaca proposed a certain quantitative description of random fuzzy subsets $FX \subseteq U$ in U, thus laying the theoretical basis for measuring fuzziness.

Definition 5.9 Suppose the fuzzy set $FX \subseteq U$, if the mapping d: $U \rightarrow [0,1]$ meets the following conditions:

1. Only if FX is a classical set, $d(FX) = 0$.
2. Only if $\mu_{FX}(x) = 0.5$, make $d(FX)$ take the biggest value.
3. If $FX \subseteq U$, $FY \subseteq U$, and $\mu_{FX}(x) \leq \mu_{FY}(x) \geq 0.5$, or $\mu_{FX}(x) \geq \mu_{FY}(x) \leq 0.5$, then $d(FX) \leq d(FY)$.
4. $d(FX) = d(FX)^C$.
5. $d(FX \cup FY) = d(FX) + d(FY) - d(FX \cap FY)$.

Then, $d(FX)$ is called the fuzzy degree of FX.

The meanings of the five conditions in the definition of fuzziness are as follows: condition (1) means the fuzzy degree of the classical set is 0; condition (2) means when the membership degree is 0.5, the fuzzy degree reaches the highest; condition (3) means the closer to 0.5, the fuzzier it is; condition (4) means the fuzzy roughness of the fuzzy set FX and its complementary set $d(FX)^C$ is same; and condition (5) requires that the operation formulas of the fuzzy degree and the probability of coincident events in the probability theory be the samc.

Generally, there are two ways of measuring fuzziness: the distance measurement and the entropy measurement.

The common distance measurements are the Hamming distance measurement and the Euclid measurement.

Suppose $FX \subseteq U$, $FY \subseteq U$, $D_H(FX, FY) = 1/n \sum_{i=1}^{n} |\mu_{FX}(x_i) - \mu_{FY}(x_i)|$ are two fuzzy sets in U, $x_i \in U$, $i = 1, 2, \ldots, n$, then $D_H(FX, FY) = 1/n \sum_{i=1}^{n} |\mu_{FX}(x_i) - \mu_{FY}(x_i)|$ is called the Hamming distance between the fuzzy sets FX and FY; $D_E(FX, FY) = 1/\sqrt{n} \sum_{i=1}^{n} \left(\mu_{FX}(x_i) - \mu_{FY}(x_i)^2\right)^{1/2}$ is regarded as the Euclid Distance between FX and FY.

If we consider only one fuzzy set $FX \subseteq U$, then how to measure its fuzziness? It is necessary to define a common set that is the closest to FX, denoted by $\underline{FX}$. Its membership function is its characteristic function:

$$\mu_{\underline{FX}}(x) = \begin{cases} 1 & \text{if } \mu_{FX} \geq 0.5 \\ 0 & \text{if } \mu_{FX} < 0.5, \end{cases}$$

then $D_H(FX) = 2/n \sum_{i=1}^{n} |\mu_{FX}(x_i) - \mu_{\underline{FY}}(x_i)|$ is called the Hamming fuzzy measurement of fuzzy set FX, while $D_E(FX) = 2/\sqrt{n}\left(\sum_{i=1}^{n} (\mu_{FX}(x_i) - \mu_{\underline{FX}}(x_i))^2\right)^{1/2}$ is called the Euclid fuzzy measurement of fuzzy set FX.

"Entropy" is originally a thermodynamic term, which is used to measure the irregularity of the molecular motion, meanwhile entropy could be used to measure the residual information in the probability theory, and also the size of fuzziness of a fuzzy set in the fuzzy set theory.

The fuzzy entropy is defined as

$$H(FX) = \frac{1}{n \ln 2} \sum_{i=1}^{n} ((-\mu_{FX}(x)) \ln \mu_{FX}(x_i) - \mu_{FX^C}(x_i) \ln \mu_{FX^C}(x_i))$$

Example 5.5 Suppose $U = \{x_1, x_2, x_3, x_4, x_5\}$ and $FX = \frac{0.2}{x_1} + \frac{0.4}{x_2} + \frac{0.8}{x_3} + \frac{0.9}{x_4} + \frac{0.6}{x_5}$, then calculate $|FX|$, $D_E(FX)$, $D_H(FX)$, $H(FX)$.

Solution: $|FX| = \sum_{i=1}^{5} \mu_{FX}(x_i) = 0.2 + 0.4 + 0.8 + 0.9 + 0.6 = 2.9$

$$D_H(FX) = \frac{2}{5}(|0.2-0| + |0.4-0| + |0.8-1| + |0.9-1| + |0.6-1|) = 0.52$$

$$D_E(FX) = \frac{2}{\sqrt{5}}((0.2-0)^2 + (0.4-0)^2 + (0.8-1)^2 + (0.9-1)^2 + (0.6-1)^2) = 0.37$$

$$\begin{aligned} H(FX) = \frac{1}{5 \times \ln 2} & ((-0.2 \times \ln 0.2 - 0.8 \times \ln 0.8) + (-0.4 \times \ln 0.4 - 0.6 \times \ln 0.6) \\ & + (-0.8 \times \ln 0.8 - 0.2 \times \ln 0.2) + (-0.9 \times \ln 0.9 - 0.1 \times \ln 0.1) \\ & + (-0.6 \ln 0.6 - 0.4 \ln 0.4)) = 0.77 \end{aligned}$$

5.2 Rough Fuzzy Set and Fuzzy Rough Set

5.2.1 Rough Fuzzy Set

When the knowledge modules in a knowledge base are all crisp concepts while the similar concepts or output classes have ill definition of the boundary, it appears the roughness and fuzziness simultaneously are due to the indiscernible relations of the input pattern sets and the fuzziness in the output classes. To simulate a situation of this type, Dubios introduced the concept of rough fuzzy sets. It is an extension of a rough set approximately deduced from a fuzzy set in a crisp and approximate space. In the rough fuzzy set, the output class is fuzzy. The definition of the upper approximation and the lower approximation of the following rough fuzzy set *FX* is given according to its basic theory.

Definition 5.10 Suppose $FS = (U, A, V, f)$ is a knowledge representation system, and I is defined as an equivalence relation on U, while FX is a fuzzy set on U, $P \subseteq A$, $FX \subseteq U$, then a pair of the lower approximation $\underline{apr}_p(FX)$ and the upper approximation $\overline{apr}_P(FX)$ of the knowledge representation system FS about FX is described, respectively, as

$$\underline{apr}_p(FX) = \inf\{x \in I(x) : \mu_{FX}(x)\} \tag{5.5}$$

$$\overline{apr}_p(FX) = \sup\{x \in I(x) : \mu_{FX}(x)\} \tag{5.6}$$

where $\mu_{FX}(x)$ is the membership degree that x belongs to FX, and the upper approximation and the lower approximation of the fuzzy set are still fuzzy sets. $\underline{apr}_p(FX)$ could be understood as the membership degree that the object definitely belongs to the fuzzy set FX, while $\overline{apr}_P(FX)$ could be understood as the membership degree that the object possibly belongs to the fuzzy set. Obviously, in crisp events, that is, when $\mu_{FX}(x) = 1$ or $\mu_{FX}(x) = 0$, the above definition is identical with the upper approximation and the lower approximation of the standard rough set.

Example 5.6 Suppose $U = \{n_1, n_2, n_3, n_4, n_5, n_6, n_7, n_8\}$ is a group consisted of eight students studied. They are divided into three equivalent classes, denoted by $U/I = \{N, M\}$, $U/I = \{\{n_1, n_6\}, \{n_2, n_7, n_8\}, \{n_3, n_4, n_5\}\}$. Suppose the fuzzy set FX expresses the fuzzy concept "height," and its membership function is $\mu_{FX}(x) = \{n_1/0.6, n_2/0.4, n_3/0.4, n_4/0.7, n_5/0.6, n_6/0.8, n_7/1, n_8/0.9\}$, calculate the upper approximation, the lower approximation, and the classification accuracy of the rough fuzzy set FX.

Solution: The upper approximation and the lower approximation of FX is described, respectively, as follows:

$$\underline{apr}_P(FX)=\left\{\frac{n_1}{0.6},\frac{n_2}{0.4},\frac{n_3}{0.4},\frac{n_4}{0.4},\frac{n_5}{0.4},\frac{n_6}{0.6},\frac{n_7}{0.4},\frac{n_8}{0.4}\right\}$$

$$\overline{apr}_P(FX)=\left\{\frac{n_1}{0.8},\frac{n_2}{1},\frac{n_3}{0.7},\frac{n_4}{0.7},\frac{n_5}{0.7},\frac{n_6}{0.8},\frac{n_7}{1},\frac{n_8}{1}\right\}$$

The classification accuracy of the rough fuzzy set *FX* is

$$\alpha_P(FX)=\frac{|\underline{apr}_P(FX)|}{|\overline{apr}_P(FX)|}=\frac{0.6+0.4+0.4+0.4+0.4+0.6+0.4+0.4}{0.8+1+0.7+0.7+0.7+0.8+1+1=1}$$

$$=\frac{3.6}{6.7}=0.5337$$

5.2.2 Fuzzy Rough Set

When the equivalence relation is not crisp, a rough fuzzy set could be extended to a fuzzy rough set. According to the basic thoughts of Dubois and Prade, the following definition of the fuzzy rough set is given.

Definition 5.11 $FS = (U, R)$ is a fuzzy approximate space. Where U is a nonempty universe, and R is an approximate relation on U, $x \in U$, $y \in U$, $P \subseteq A$, $FX \subseteq U$, the input class $F_1, F_2, \ldots, F_n$ is a fuzzy cluster originated from a fuzzy equivalence relation, and n is the number of the cluster. Every F_i is a fuzzy equivalent class. Thus, a pair of the upper approximation and the lower approximation of the fuzzy rough set FX on U is

$$\underline{R}_P(FX)=\inf\{\max\{1-\mu_{F_I}(x),\mu_{FX}(x)\}\} \tag{5.7}$$

$$\overline{R}_P(FX)=\sup\{\min\{\mu_{F_i}(x),\mu_{FX}(x)\}\} \tag{5.8}$$

Obviously, when the equivalence relation is clear, a fuzzy rough set will degrade into a rough fuzzy set. When all the fuzzy equivalence relations are clear, it will further degrade into a classical rough set.

5.3 Variable Precision Rough Fuzzy Sets

Variable precision rough fuzzy set is an extension of the rough set when the output class is a fuzzy set, while the approximation space is clear and decision attribute value is fuzzy in variable precision rough fuzzy set.

5.3.1 Rough Membership Function Based on λ-Cut Set

Definition 5.12 Suppose $FS = (U, A, V, f)$ is a decision table, $A = C \cup D$, $C \cap D = \phi$, and I is an equivalence relation defined on U, while the output set $FX \subseteq U$ is a fuzzy set on U, then the table is called rough fuzzy decision table.

Because the λ-cut set of fuzzy set FX is a common set, for any common set FX_λ in the output fuzzy set $FX \subseteq U$, and rough membership function of FX_λ can be defined as follows with the application of the rough membership function:

$$\mu_{FX_\lambda}(x) = \frac{|(I(x) \cap FX_\lambda|}{|I(x)|} \tag{5.9}$$

where

$\mu_{FX_\lambda}(x)$ denotes the degree that x belongs to the set FX_λ

$|FX_\lambda|$ denotes the sum of the membership degree of all elements in set FX_λ

Proposition 5.2 For any output subset $FX_\lambda \subseteq U$ in rough fuzzy decision table $FS = (U, A, V, f)$, there exists $0 \le \mu_{FX_\lambda}(x) \le 1$.

Proof: $\phi \subseteq I(x) \cap FX_\lambda \subseteq I(x)$, thus we can infer $0 \le (|I(x) \cap FX_\lambda|/|I(x)|) \le 1$, that is to say, $0 \le \mu_{FX_\lambda}(x) \le 1$.

Proposition 5.3 $\mu_{U-FX_\lambda}(x) = 1 - \mu_{FX_\lambda}(x)$

Proof: $\mu_{U-FX_\lambda}(x) = \frac{|(I(x) \cap (U - FX_\lambda)|}{|I(x)|} = \frac{|I(x) \cap U|}{|I(x)|} - \frac{|I(x) \cap FX_\lambda|}{|I(x)|} = 1 - \mu_{FX_\lambda}(x)$

Proposition 5.4 For any two output subsets $X_\lambda \subseteq U$ and $Y_\lambda \subseteq U$ in rough fuzzy decision table $FS = (U, A, V, f)$, if $X_\lambda \subseteq Y_\lambda$, then $\mu_{X_\lambda}(x) \le \mu_{Y_\lambda}(x)$.

Proof: For $\forall x \in U$, due to $X_\lambda \subseteq Y_\lambda$, therefore $|I(x) \cap X_\lambda| \le |I(x) \cap Y_\lambda|$, thus we can obtain $(|I(x) \cap X_\lambda|/|I(x)|) \le (|I(x) \cap Y_\lambda|/|I(x)|)$, that is to say, $\mu_{X_\lambda}(x) \le \mu_{Y_\lambda}(x)$.

Proposition 5.5 If any two output subsets $X_\lambda \subseteq U$ and $Y_\lambda \subseteq U$ in rough fuzzy decision table, the following relations exist:

1. $\mu_{X_\lambda \cup Y_\lambda}(x) \ge \max\{\mu_{X_\lambda}(x), \mu_{Y_\lambda}(x)\}$
2. $\mu_{X_\lambda \cap Y_\lambda}(x) \le \min\{\mu_{X_\lambda}(x), \mu_{Y_\lambda}(x)\}$

Proof

1. For $\forall x \in U$,

$$\mu_{X_\lambda \cup Y_\lambda}(x) = \frac{|I(x) \cap (X_\lambda \cup Y_\lambda)|}{|I(x)|} = \frac{|(I(x) \cap X_\lambda) \cup (I(x) \cap Y_\lambda)|}{|I(x)|}$$

$$\geq \frac{\max\{|I(x) \cap X_\lambda|, |I(x) \cap Y_\lambda|\}}{|I(x)|} = \max\left\{\frac{|I(x) \cap X_\lambda|}{|I(x)|}, \frac{|I(x) \cap Y_\lambda|}{|I(x)|}\right\}$$

$$= \max\{\mu_{X_\lambda}, \mu_{Y_\lambda}\}$$

Therefore, $\mu_{X_\lambda \cup Y_\lambda}(x) \geq \max\{\mu_{X_\lambda}(x), \mu_{Y_\lambda}(x)\}$.

2. For $\forall x \in U$, then

$$\mu_{X_\lambda \cap Y_\lambda}(x) = \frac{|I(x) \cap (X_\lambda \cap Y_\lambda)|}{|I(x)|} = \frac{|(I(x) \cap X_\lambda) \cap (I(x) \cap Y_\lambda)|}{|I(x)|}$$

$$\leq \frac{\min\{|I(x) \cap X_\lambda|, |I(x) \cap Y_\lambda|\}}{|I(x)|} = \min\left\{\frac{|I(x) \cap X_\lambda|}{|I(x)|}, \frac{|I(x) \cap Y_\lambda|}{|I(x)|}\right\}$$

$$= \min\{\mu_{X_\lambda}(x), \mu_{Y_\lambda}(x)\}$$

therefore, $\mu_{X_\lambda \cap Y_\lambda}(x) \leq \min\{\mu_{X_\lambda}(x), \mu_{Y_\lambda}(x)\}$.

5.3.2 The Rough Approximation of Variable Precision Rough Fuzzy Set

To deduce non-strong decision rules used in probabilistic decision evaluation, we should consider the degree of overlapping. Let confidence $0.5 < \beta \leq 1$ and $P \subseteq C$, then β-lower approximation based on λ-cut set in fuzzy set $FX \subseteq U$ is defined as follows:

$$\underline{apr}_\beta^P(FX) = \cup\{x \in U : \mu_{FX_\lambda}(x) \geq \beta\} \tag{5.10}$$

$$\overline{apr}_\beta^P(FX) = \cup\{x \in U : \mu_{FX_\lambda}(x) \geq 1 - \beta\} \tag{5.11}$$

When two different decision subsets in the universe U have nonempty overlapping, β-lower approximation of FX can be interpreted as the union of equivalence class of $\mu_{FX_\lambda}(x) \geq \beta$, while β-upper approximation of FX can be interpreted as the union of equivalence class of $\mu_{FX_\lambda}(x) < \beta$.

β-Lower approximation of FX measures the precision that condition class is allotted to fuzzy decision class, and is similar to the correct classification proportion in variable precision rough set. Obviously, when the output set FX is clear, due to $\underline{apr}_{\beta}^{P}(FX) = \bigcup\{x \in U \mid (|I(x) \cap FX| / |I(x)|) = 1\} = \bigcup\{x \in U \mid I(x) \subseteq FX\}$, then β-lower approximation of FX is degraded into the lower approximation of rough set model, and $\overline{apr}_{\beta}^{P}(FX) = \bigcup\{x \in U \mid (|I(x) \cap FX| / |I(x)|) > 0\} = \bigcup\{x \in U \mid I(x) \cap FX \neq \phi$, then β-upper approximation of FX is degraded into the upper approximation of rough set model.

When the output set FX is clear and $\beta = 1$, variable precision rough fuzzy set model is degraded into rough set model.

5.3.3 The Approximate Quality and Approximate Reduct of Variable Precision Rough Fuzzy Set

Suppose $0.5 < \beta \leq 1$, the approximate quality of variable precision rough fuzzy set is defined as follows:

$$\gamma_{P}^{\beta}(D) = \frac{\left|\bigcup\left\{\dfrac{|I(x) \cap FX_{\lambda}|}{|I(x)|} \geq \beta\right\}\right|}{|U|} \tag{5.12}$$

The classification quality $\gamma_{P}^{\beta}(D)$ measures the proportion of rightly classified knowledge in available knowledge in variable precision rough fuzzy decision table at a given confident threshold value β in the universe.

An approximate reduct $red_{P}^{\beta}(C,D)$ of variable precision rough fuzzy set is defined as the minimal subset $P \subseteq C$ that satisfies $\gamma_{P}^{\beta}(D) = \gamma_{C}^{\beta}(D)$ at a given confident threshold value β.

One rough fuzzy decision table may have more than one reduct, and the intersection of all the reduct is called core. A core may also be an empty set.

5.3.4 The Probabilistic Decision Rules Acquisition of Rough Fuzzy Decision Table

Probabilistic decision rules can be derived from the approximate reduction $red_{P}^{\beta}(C,D)$ in rough fuzzy decision table. The sets of condition elements in the universe are called condition classes in FS denoted by $C_i (i = 1, 2, \ldots, k)$; the sets of decision elements in the universe are called decision classes in FS denoted by FX_j and $C_i \cap FX_j = \phi$, then the rules $r: CON_C(C_i) \xrightarrow{\beta} DEC_D(FX_j)$ are referred to probabilistic rules of (C, D) denoted by $\{r_{ij}\}$, and confident degree is β. The syntax of the rules is as follows:

If $f(x, q_1) = r_{q1} \wedge f(x, q_2) = r_{q2} \wedge \ldots \wedge f(x, q_p) = r_{qp}$, then $x \in FX_j$ with β where $\{q_1, q_2, \ldots, q_p\} \subseteq C$, $(r_{q1}, r_{q2}, \ldots, r_{qp}\} \in V_{q1} \times V_{q2} \times \cdots \times V_{qp}$, $j \in \{1, 2, \ldots, m\}$ denotes a certain decision class.

5.3.5 Algorithm Design

Improved rapid reduct algorithm is applied to design a reduct algorithm to acquire probabilistic decision rules from rough fuzzy decision table. Reduct algorithm is described as follows:

Input: The set of conditional attributes C and the set decision attributes D.
Output: The minimal reduct of conditional attributes R.
Step 1: Indiscernibility class $I(x)$ generated from the set of conditional attributes.
Step 2: Decision class FX produced from the set decision attributes.
Step 3: Calculate λ-cut set of every decision class and form fuzzy equivalence class FX_λ.
Step 4: Calculate membership function $\mu_{FX_\lambda}(x) = |I(x) \cap FX_\lambda| / |I(x)|$ of every decision class.
Step 5: Calculate the classification quality $\gamma_C^\alpha(D) = |\cup\{|I(x) \cap FX_\lambda| / |I(x)| \geq \beta\}| / |U|$ for all sets of conditional attributes C according to the setting confident threshold value $0.5 < \beta \leq 1$.
Step 6: $R = C$.
Step 7: $S = R$.
Step 8: For every conditional attribute $c_i \in C$, calculate $\gamma_{R-\{c_i\}}^\beta(D)$.
Step 9: If $\gamma_{R-\{c_i\}}^\beta(D) = \gamma_S^\beta(D)$.
Step 10: Then $S = R - \{c_i\}$, $R_1 = R$, $R = S$.
Step 11: Execute repeatedly from Step 7 to Step 10 until $\gamma_R^\beta(D) < \gamma_C^\beta(D)$ or $R = R_1$.

Example 5.7 A discretized information decision table for sales volume is given in Table 5.1, where c_1 represents quality of goods, c_2 represents price of goods, c_3 represents promotion, c_4 represents service, and d represents sales volume of goods. A manager's judgment on the level of sales volume for 13 objects applied is shown in Table 5.2, where d_1 denotes that sales volume of goods is high, d_2 denotes that sales volume of goods is medium, and d_3 denotes that sales volume of goods is low. If his or her decision is clear, he or she sets one of the sales levels as 1 and others as 0. For example, the object n_2 makes such a decision: $d_1 = 1$, $d_2 = 0$, $d_3 = 0$. If his or her decision is vague, he or she prefers a possible decision by adopting a fuzzy membership value. For example, the object n_1 makes a decision $d_1 = 0.89$, $d_2 = 0.11$, $d_3 = 0$, which reflects the fuzziness of his judgment on the level of sales volume in data base.

According to the algorithm designed in this section, we can get the following equivalence class by dividing the universe U based on the conditional attribute set C:

$$\frac{U}{C} = \{X_1, X_2, X_3, X_4, X_5, X_6, X_7\}$$

Table 5.1 An Information Decision Table for Sales Volume

U	*Goods Attribute (C)*				*Decision Attribute (D) d*
	c_1	c_2	c_3	c_4	
n_1	Good	Low	Without	Good	High
n_2	Good	Low	With	Good	High
n_3	Good	Low	Without	Good	High
n_4	Good	Low	Without	Good	High
n_5	Good	Low	Without	Good	High
n_6	Good	High	With	Medium	Medium
n_7	Good	High	Without	Medium	Low
n_8	Medium	High	With	Good	Low
n_9	Medium	High	With	Good	Medium
n_{10}	Medium	Low	With	Good	Medium
n_{11}	Medium	Low	Without	Medium	Low
n_{12}	Good	Low	Without	Good	Medium
n_{13}	Good	Low	Without	Good	High

where $X_1 = \{n_1, n_3, n_4, n_5, n_{12}, n_{13}\}$, $X_2 = \{n_2\}$, $X_3 = \{n_6\}$, $X_4 = \{n_7\}$, $X_5 = \{n_8, n_9\}$, $X_6 = \{n_{10}\}$, $X_7 = \{n_{11}\}$.

We can get the following decision class by dividing the universe based on the decision attribute set D:

$$\frac{U}{D} = \{FH, FM, FL\}$$

where $FH = \{n_1, n_2, n_3, n_4, n_5, n_{13}\}$, $FM = \{n_6, n_9, n_{10}, n_{12}\}$, $FL = \{n_7, n_8, n_{11}\}$.

Suppose $\lambda = \min\{\mu_1(x), \mu_2(x), \ldots, \mu_t(x)\}$, where t denotes the number of elements contained in a decision class.

Calculate fuzzy membership functions of every decision class FX_λ. For example, the fuzzy membership function of equivalence class X_1 for FH_λ is as follows:

$$\mu_{FH_\lambda}(x) = \frac{|I(x) \cap FX_\lambda|}{|I(x)|} = \frac{|X_1 \cap FH_\lambda|}{|X_1|} = \frac{0.89 + 0.86 + 0.88 + 0.83 + 0.9}{6} = 72.7 \text{ percent}$$

Table 5.2 Manager's Fuzzy Partition Table of Sales Volume

U	Fuzzy Decision Membership Value			Practical Decision Value d
	d_1	d_2	d_3	
n_1	0.89	0.11	0	High
n_2	1	0	0	High
n_3	0.86	0.14	0	High
n_4	0.88	0.12	0	High
n_5	0.83	0.17	0	High
n_6	0	0.80	0.20	Medium
n_7	0	0.13	0.87	Low
n_8	0	0.31	0.69	Low
n_9	0	0.54	0.46	Medium
n_{10}	0	0.81	0.19	Medium
n_{11}	0	0	1	Low
n_{12}	0.20	0.80	0	Medium
n_{13}	0.90	0.10	0	High

Similarly, other fuzzy membership function values can be acquired. Let confident threshold value $\alpha = 70$ percent, we can get the classification quality for all sets of conditional attributes C as follows:

$$\gamma_C^\alpha(D) = \frac{\left| \bigcup \left\{ \frac{|I(x) \cap FX_\lambda|}{|I(x)|} \geq \alpha \right\} \right|}{|U|} = \frac{11}{13} = 84.6 \text{ percent}$$

For every conditional attribute $c_i \in C$, calculate $\gamma_{C-\{c_i\}}^\alpha(D)$, and repeat the process, and finally we can get a minimal β-reduct $\{c_1, c_2, c_3\}$. The minimal probabilistic decision rules are derived according to β-reduct $\{c_1, c_2, c_3\}$ and shown in Table 5.3.

Table 5.3 Probabilistic Decision Rules for β-Reduct $\{c_1, c_2, c_3\}$

Rules	Support	Confidence (Percent)
$c_1 = \text{good} \wedge c_2 = \text{low} \wedge c_3 = \text{without} \xrightarrow{72.7\text{ percent}} d \in FH$	6	72.7
$c_1 = \text{good} \wedge c_2 = \text{low} \wedge c_3 = \text{with} \xrightarrow{100\text{ percent}} d \in FH$	1	100
$c_1 = \text{good} \wedge c_2 = \text{high} \wedge c_3 = \text{with} \xrightarrow{80\text{ percent}} d \in FM$	1	80
$c_1 = \text{good} \wedge c_2 = \text{high} \wedge c_3 = \text{without} \xrightarrow{87\text{ percent}} d \in FL$	1	87
$c_1 = \text{medium} \wedge c_2 = \text{low} \wedge c_3 = \text{with} \xrightarrow{81\text{ percent}} d \in FM$	1	81
$c_1 = \text{medium} \wedge c_2 = \text{low} \wedge c_3 = \text{without} \xrightarrow{100\text{ percent}} d \in FH$	1	100

5.4 Variable Precision Fuzzy Rough Set

When the knowledge in knowledge base is fuzzy while the approximate concept is clear, fuzzy variable precision rough set can be applied to solve the problems of probabilistic decision.

5.4.1 Fuzzy Equivalence Relation

Definition 5.13 Suppose $FS = (U, R)$ is a fuzzy approximation space, U is a nonempty universe, R is the fuzzy relation on U, and the membership function of R is denoted by μ_R if R satisfies all of the following three properties:

1. Reflexivity: $\mu_R(x, x) = 1, \forall x \in U$
2. Symmetry: $\mu_R(x, y) = \mu_R(y, x), \forall \in x, y \in U$
3. Transitivity: $\forall \in x, y, z \in U, \forall \lambda \in [0, 1]$, if $(y, z) \geq \lambda$, $\mu_R(x, y) \geq \lambda$, then $\mu_R(x, z) \geq \lambda$

then R is called the fuzzy equivalence relation on U.

Proposition 5.6 $FS = (U, R)$ is a fuzzy approximation space, where R denotes a fuzzy equivalence relation on U. With a λ-cut set on fuzzy equivalence relation R, for $0 \leq \lambda \leq 1$, every λ horizontal cut set is a common equivalence relation, denoted by R_λ.

Proof

1. *Reflexivity*: For $\forall x \in U$, $\forall y \in U$, because $\mu_R(x, x) = 1$, therefore for $\forall y \in [0, 1]$, there exists $\mu_R(x, x) \geq \lambda$, thus $\mu_{R_\lambda}(x, x) = 1$, that is to say, R_λ possesses reflexivity.
2. *Symmetry*: Because $\mu_R(x, y) = \mu_R(y, x)$, for $\forall\lambda \in [0, 1]$, when $\mu_R(x, y) = \mu_R(y, x) \geq \lambda$, then $\mu_{R_\lambda}(x, y) = \mu_{R_\lambda}(y, x) = 1$; when $\mu_R(x, y) = \mu_R(y, x) < \lambda$, $\mu_{R_\lambda}(x, y) = \mu_{R_\lambda}(y, x) = 0$. Hence for $\forall\lambda \in [0, 1]$, there always exists $\mu_{R_\lambda}(x, y) = \mu_{R_\lambda}(y, x)$. That is to say, R_λ possesses symmetry.
3. *Transitivity*: Because R possesses transitivity, take some arbitrary element $(x, y) \in R_\lambda$, $(y, z) \in R_\lambda$, then $R(x, y) \geq \lambda$, $R(y, z) \geq \lambda$. So for $\forall t \in U$, there exists $R(x, z) \geq R^2(x, z) = \vee (R(x, t) \wedge R(t, z)) \geq R(x, y) \wedge R(y, z) \geq \lambda$, therefore, $(x, z) \in R_\lambda$. That is to say, R_λ possesses transitivity.

Hence, for $\forall\lambda \in [0, 1]$, R_λ is common equivalence relation. These equivalence relations R_λ can divide the universe U into several fuzzy equivalence classes that may fold.

5.4.2 Variable Precision Fuzzy Rough Model

Definition 5.14 Suppose $FS = (U, R)$ is a fuzzy approximation space, U is nonempty universe, and R is the fuzzy equivalence relation of U, while R_λ is the λ-cut set of R and $X \subseteq U$. Fuzzy set F_X based on fuzzy equivalence relation is defined as follows:

$$F_X = \left\{(x, \mu_{F_X}(x)) : x \in U, \mu_{F_X}(x) = \frac{|R_\lambda(x) \cap X|}{|R_\lambda(x)|}\right\} \tag{5.13}$$

where

$\mu_{F_X}(x)$ is the membership function of fuzzy set F_X and denotes the degree that element x belongs to fuzzy set F_X

$R_\lambda(x)$ is the fuzzy equivalence class containing element x

$R_\lambda(x)$ is the sum of the membership degree of elements contained in fuzzy equivalence class $R_\lambda(x)$

For every λ-cut set in fuzzy equivalence relation R, variable precision rough set model can be applied, thus β-upper approximation and β-lower approximation of variable precision rough set can be derived from λ horizontal fuzzy equivalence relation for a specified subset $X \subseteq U$. Every approximation is still a fuzzy set among them.

Definition 5.15 $FS = (U, A, V, f)$ is a fuzzy decision table, U is nonempty universe, and $A = C \cup D$, $P \subseteq C$ is the fuzzy equivalence relation on U, while R is the fuzzy equivalence relation on U, R_λ is the λ-cut set of R, $x \in U$, $X \subseteq U$, $0.5 < \beta \leq 1$. Then β-upper approximation and β-lower approximation of X are defined, respectively, by

$$\underline{R}_P^\beta(\mathrm{X}) = \bigcup \sup\{x \in U : \mu_{F_X}(x) \geq \beta\} \tag{5.14}$$

$$\overline{R}_P^\beta(\mathrm{X}) = \bigcup \sup\{x \in U : \mu_{F_X}(x) \geq 1 - \beta\} \tag{5.15}$$

When two different subsets have nonempty fold in the universe U, the lower approximation of X can be interpreted as the union of the equivalence classes that membership function $\mu_{F_X}(x)$ is not below the supremum of β, while the upper approximation of X can be interpreted as the union of the equivalence classes membership function $\mu_{F_X}(x)$ above the infimum of $1 - \beta$. Obviously, when R is equivalence relation, it is degraded into variable precision rough set model. Due to $\beta > 1 - \beta$, the key property $\underline{R}_P^\beta(X) \subseteq \overline{R}_P^\beta(X)$ still holds.

Proposition 5.7 For any subset $X \subseteq U$ in fuzzy approximation space $FS = (U, R)$, then $0 \leq \mu_{F_X}(x) \leq 1$.

Proof: $\phi \subseteq R_\lambda(x) \cap X \subseteq R_\lambda(x)$, thus we can infer $0 \leq (|R_\lambda(x) \cap X|/|R_\lambda(x)|) \leq 1$, that is to say, $0 \leq \mu_{F_X}(x) \leq 1$.

Proposition 5.8 For any two subsets $X \subseteq U$ and $Y \subseteq U$ in fuzzy approximation space $FS = (U, R)$, R is the fuzzy equivalence relation on U, R_λ is the λ-cut set of R, then

1. $F_{X \cup Y} \supseteq F_X \cup F_Y$
2. If $X \subseteq Y$ or $Y \subseteq X$, then $F_{X \cup Y} = F_X \cup F_Y$
3. $F_{X \cap Y} \subseteq F_X \cap F_Y$
4. If $X \subseteq Y$ or $Y \subseteq X$, then $F_{X \cap Y} \subseteq F_X \cap F_Y$

Proof

1. For $\forall x \in U$,

$$\mu_{F_{X \cup Y}}(x) = \frac{|R_\lambda(x) \cap (X \cup Y)|}{|R_\lambda(x)|} = \frac{|(R_\lambda(x) \cap X) \cup (R_\lambda(x) \cap Y)|}{|R_\lambda(x)|}$$

$$\geq \frac{\max\{|R_\lambda(x) \cap X|, |R_\lambda(x) \cap Y|\}}{|R_\lambda(x)|} = \max\left\{\frac{|R_\lambda(x) \cap X|}{|R_\lambda(x)|}, \frac{|R_\lambda(x) \cap Y|}{|R_\lambda(x)|}\right\}$$

$$= \max\left\{\mu_{F_X}, \mu_{F_Y}\right\} = \mu_{F_X \cup F_Y}$$

therefore, $F_{X \cup Y} \supseteq F_X \cup F_Y$.

2. The proof of (2) can be derived from the proof of (1)

3. $$\mu_{F_{X\cap Y}}(x)=\frac{|R_\lambda(x)\cap(X\cap Y)|}{|R_\lambda(x)|}=\frac{|(R_\lambda(x)\cap X)\cap(R_\lambda(x)\cap Y)|}{|R_\lambda(x)|}$$

$$\le\frac{\min\{|R_\lambda(x)\cap X|,|R_\lambda(x)\cap Y|\}}{|R_\lambda(x)|}=\min\left\{\frac{|R_\lambda(x)\cap X|}{|R_\lambda(x)|},\frac{R_\lambda(x)\cap Y|}{|R_\lambda(x)|}\right\}$$

$$=\min\{\mu_{F_X},\mu_{F_Y}\}=\mu_{F_X\cap F_Y}$$

therefore, $F_{X\cap Y}\subseteq F_X\cap F_Y$.

4. The proof of (4) can be derived from the proof of (3)

According to the definition of variable precision fuzzy rough set, if a subset $X\subseteq U$ and a fuzzy equivalence relation R on U are given, for confident threshold value $0.5<\beta\le 1$, variable X can be classified into the following four classes:

1. If $\underline{R}_P^\beta(X)\neq\phi$, $\overline{R}_P^\beta(X)\neq U$, then X is partly definable.
2. If $\overline{R}_P^\beta(X)\neq\phi$, $\overline{R}_P^\beta(X)=U$, then X is internally definable.
3. If $\underline{R}_P^\beta(X)=\phi$, $\overline{R}_P^\beta(X)\neq U$, then X is externally definable.
4. If $\underline{R}_P^\beta(X)=\phi$, $\overline{R}_P^\beta(X)=U$, then X is completely not definable.

Such a classification is actually a fuzzy extension of classifications in the variable precision rough set.

5.4.3 Acquisition of Probabilistic Decision Rules in Fuzzy Rough Decision Table

The sets of all fuzzy condition elements in the universe are called condition classes in FS, denoted by $F_{X_i}(i=1,2,\ldots,k)$; the sets of all decision elements in the universe are called decision classes in FS, denoted by $Y_j(j=1,2,\ldots,k)$, and $F_{X_i}\cap Y_j=\phi$, then, the rule r: $CON_C(F_{X_i})\xrightarrow{\beta}DEC_D(Y_j)$ is called probabilistic rules of (C, D) with confident degree β, denoted by $\{r_{ij}\}$. The syntax of the rules is as follows: If $f(x,q_1)=r_{q1}\wedge f(x,q_2)=r_{q2}\wedge\ldots\wedge f(x,q_p)=r_{qp}$, then $x\in Y_j$ with β where $\{q_1, q_2, \ldots, q_p\}\subseteq C$, $(r_{q1},r_{q2},\ldots,r_{qp}\}\in V_{q1}\times V_{q2}\times\cdots\times V_{qp}$, $j\in\{1,2,\ldots,m\}$ presents a certain class of given decision attribute.

5.4.4 Measure Methods of the Fuzzy Roughness for Output Classification

Fuzzy roughness has influence on the management of the uncertainty knowledge in many classification designs, and we give two quantification measurements that are average fuzzy roughness of output classification with the distance measurement of fuzzy set and the fuzzy entropy for reference.

5.4.4.1 Distance Measurement

$$D(F_X)=\frac{2}{n^{1/t}}\left[\sum_{x\in U}\left|\mu_{F_X}(x)-\mu_{\underline{F_X}}(x)\right|^t\right]^{1/t}$$

where

n means the number of objects contained

t = 1, 2,... is the natural number chosen by decision-maker the value of t depends on the type of applicable distance function

In practical application, t often takes the value 1 or 2: when t = 1, Hamming distance measurement is also called linear fuzzy measurement. Even though this measurement is convenient and easy, it needs to get absolute value during calculation, and sometimes its precision is poor; when t = 2, Euclidean distance measurement is also called binary time fuzzy exponent measurement. Euclidean distance measurement is commonly adopted when $t \to \infty$, and it is called the Chebyshev distance.

$\mu_{\underline{F_X}}(x)$ is the membership function closest to $\mu_{F_X}(x)$ without fuzzy roughness. $\mu_{F_X}(x)$ is defined by

$$\mu_{\underline{F_X}}(x)=\begin{cases}1 & \text{when } \mu_{FX}\geq 0.5\\ 0 & \text{when } \mu_{FX}<0.5\end{cases}$$

Thus, conclusions can be made from the above distance fuzzy measurement when we determine whether an input object belongs to a certain category, the fuzzy roughness of the set is 0 on the condition that there is no fuzzy roughness; if the fuzzy roughness of the input category is maximal, for example, for $\forall x \in U$, it is 0.5, then the fuzzy roughness of corresponding output category should be maximal. When the fuzzy rough membership value of input pattern is 0 or 1, the uncertainty of pattern ownership in output class will decrease. Therefore, the degree of fuzzy roughness of output category should also decrease.

5.4.4.2 Entropy Measurement

Fuzzy entropy is defined by

$$H(FX)=k\sum_{x\in U}((-\mu_{F_X}(x))\ln(\mu_{F_X}(x))-(1-\mu_{F_X}(x))\ln(1-\mu_{F_X}(x)))$$

where $k = 1/(n \ln 2)$, for $\forall x \in U$, when $\mu_{FX}(x)=1$ or $\mu_{FX}(x)=0$, fuzzy entropy is minimal, and let the minimal fuzzy entropy $H(FX) = 0$, that is, classical set is not fuzzy and rough.

The measure of fuzzy roughness is very important in evaluating the total uncertainty of output category. Apart from methods mentioned above, there are also distance measurement, weight distance measurement, approach degree defined by matrix, and other forms of fuzzy entropy. So far, we have not had a totally satisfying fuzzy measurement that is applicable to all practical problems, and the above methods aim at the analysis and comparison between different fuzzy sets so as to illustrate which fuzzy rough set is clearer with the various quantity indexes of different measurements.

Example 5.8 Suppose the fuzzy membership function of every attribute given by experts is as depicted in Figure 5.3.

According to the membership function, we can convert the measured data of every object to the membership degree. For example, the SP attribute value 122 of the object n_1 is converted to the membership degree 0.9/*N*, while DP attribute value 80 is converted to the membership degree 0.9/*N*, thus a fuzzy decision table $FS = \{U, A, V, f\}$ can be obtained and shown in Table 5.4, where $U = \{n_1, n_2, \ldots, n_7\}$ denotes a data set that is made up of seven fuzzy objects, and C = {SP(Systolic Pressure) and DP(Diastolic Pressure)} are conditional attribute sets made up of two fuzzy conditional attributes, while *D* = {BP(Blood Pressure)} is a decision attribute set that is made up of one decision attribute.

The equivalence class generated from decision attribute BP is as follows:

$$\frac{U}{D} = \{X_N, X_H, X_L\}$$

where $X_N = \{n_1, n_3\}$, $X_H = \{n_2, n_5, n_6\}$, $X_L = \{n_4, n_7\}$.

According to fuzzy conditional attribute SP, and DP based on that, a λ-cut set can be applied to fuzzy equivalence relation *R* in fuzzy conditional attribute, fuzzy equivalence class is as follows:

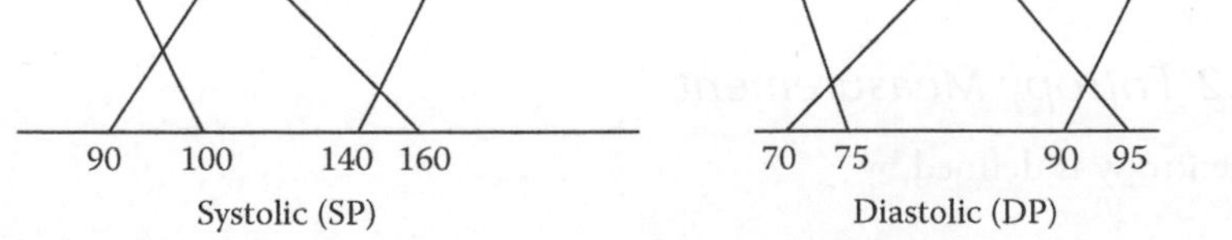

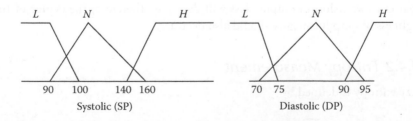

Figure 5.3 Fuzzy membership function of every attribute given by experts.

Table 5.4   Fuzzy Decision Table

U	Conditional Attributes		Decision Attributes BP
	SP	*DP*	
n_1	0.9/N	0.9/N	N
n_2	0.1/N + 0.75/H	0.4/N	H
n_3	0.85/N	0.3/N + 0.4/H	N
n_4	1/L	1/L	L
n_5	1/H	0.16/N + 0.6/H	H
n_6	0.4/N	1/H	H
n_7	0.5/L + 0.1/N	0.4/N	L

$$\frac{U}{D} = \left\{ (\{n_1,n_2,n_3,n_7\},0.1), (\{n_2,n_5\},0.75), (\{n_4,n_7\},0.5), (\{n_3,n_5,n_6\},0.4), (\{n_1,n_2,n_3,n_7\},0.3) \right\}.$$

We can obtain the membership function $\sup\{x \in U : \mu_{F_X}(x)\}$ of every decision class with fuzzy equivalence class. For instance, we can get the following membership function for decision class X_N:

$$\sup\left\{ \frac{|(R_\lambda(x)) \cap X|}{|R_\lambda(x)|} \right\} = \sup\left\{ \frac{|\{n_1,n_2,n_3,n_7\} \cap \{n_1,n_3\}|}{|\{n_1,n_2,n_3,n_7\}|} \right\}$$

$$= \sup\left\{ \frac{0.9+0.85}{0.9+0.1+0.85+0.1}, \frac{0.9+0.3}{0.9+0.4+0.3+0.4} \right\} = 0.897$$

Table 5.5   Approximation of Decision Class and the Fuzzy Roughness Measure

Decision Class	β-Lower Approximation	β-Upper Approximation	Linear Fuzzy Degree	Binary Time Fuzzy Exponent	Fuzzy Entropy
X_N	{({n_1, n_2, n_3, n_7}, 0.1)}	{({n_1, n_2, n_3, n_7}, 0.1)}	0.29	0.30	0.58
X_H	{({n_2, n_5}, 0.75)}	{({n_2, n_5}, 0.75), ({n_3, n_5, n_6}, 0.4)}	0.23	0.27	0.48
X_L	{({n_4, n_7}, 0.5)}	{({n_4, n_7}, 0.5)}	0.10	0.10	0.29

Table 5.6 Minimal Probabilistic Decision Rules

Rules	*Degree of Confidence (Percent)*
$SP = N \wedge DP = N \xrightarrow{\text{89.7 percent}} BP = N$	89.7
$SP = H \xrightarrow{\text{100 percent}} BP = H$	100
$SP = L \xrightarrow{\text{100 percent}} BP = L$	100

Similarly, $\sup\{x \in U : \mu_{F_X}(x)\}$ of other decision class can be acquired. Suppose confidence level $\beta = 0.85$, we can get β-lower approximation, β-upper approximation, linear fuzzy degree, binary time fuzzy exponent, and fuzzy entropy of subsets X_N, X_H, X_L as shown in Table 5.5.

From Table 5.5, we can know that the uncertainty of the knowledge in X_N is maximal, and the uncertainty of the knowledge in X_H is medium while that is minimal in X_L. The minimal probabilistic decision rules of β-lower approximation can be generated from Table 5.5 as shown in Table 5.6.

5.5 Summary

One of the main research topics in knowledge universe is to simulate inaccurate and incomplete information, and many available methods are some extensions based on classical set theories, such as fuzzy set theory and rough set theory. The concepts of fuzzy set theory and rough set theory are not rivaling ones but two different mathematical tools with different purposes. Rough set theory deals with the approximation of sets when the elements are indiscernible, while fuzzy set theory deals with fuzzy concept by allowing part membership, and these two theories are all motivated by practical needs and aim at solving practical problems. This chapter mainly introduces the approximation of fuzzy sets in crisp approximation space and the approximation of crisp sets in fuzzy approximation. When the knowledge modules in knowledge base are all clear concepts but approximated concept or output classes have the ill definition of boundaries, we can solve the decision problems with rough fuzzy set model; when the knowledge modules in knowledge base are all fuzzy concepts while the approximated concept is a clear one, we can apply fuzzy rough set model to solving such decision problems. As a result of the inner fuzziness of the decision-maker's thinking and the existence of noisy data, it is sometimes impossible to acquire decision rules but likely to acquire strong probabilistic decision rules. To deduce the probabilistic decision rules when the knowledge in knowledge base or approximated concept is clear, we have studied hybrid model between variable precision rough set and fuzzy set. For rough fuzzy decision table, we convert output fuzzy set to common set by using λ-cut set, and we extend rough membership function by setting confident threshold value β based on this and construct a variable precision rough fuzzy set model that is probabilistic decision analysis of rough fuzzy decision table; for fuzzy rough decision table, we convert fuzzy equivalence relation into equivalence relation through λ-cut set, and we construct variable precision fuzzy rough set model based on research mentioned above, and integrate variable precision rough set and fuzzy set, then develop the concept of variable precision rough set, further extend the function of variable precision rough set method, and analyze some properties among them. Because every output class is associated with corresponding fuzzy quantity, we put forward distance and entropy fuzzy measurements to measure the relative fuzziness of output class.

Chapter 6

Hybrid of Rough Set and Grey System

Grey system theory and rough set theory are two different mathematical tools that are used to deal with uncertain or incomplete information, and yet they are relevant and complementary to a certain degree. They both improve the generality of data presentation by reducing its accuracy. For example, grey system theory reduces the accuracy of data presentation through grey sequences generating, while rough set is through discrete data, which makes it possible to find models from data that may be fuzzed by too many details, and neither of them needs a priori knowledge, such as probability distributions or membership, and so on. Rough set theory researches into the rough categories of nonoverlap and rough concepts, and concentrates on the indiscernibility between objects, while grey system theory researches into grey fuzzy sets, which have "clear extension, unclear connotation," and concentrates on the uncertainty of poor information. The appropriate hybrid of the two theories can overcome the shortages of their definitions and applications and thus has more powerful functions.

6.1 The Basic Concepts and Methods of the Grey System Theory

Grey system theory was pioneered by well-known Chinese scholar, Professor Julong Deng, in 1982. It is used to deal with the problem of uncertainity in less data little sample, which is designed as greyness, thus the system of what having greyness is said to be grey system; accordingly, there are whitening systems: complete information, and black systems: devoid of information.

For example, the human body is a grey system, because most parameters of the human body, apart from some external parameters such as height, weight, body temperature, blood pressure, and so on, are unknown. Besides, in the fields of agriculture, industry, socioeconomics, and so on, there are many grey systems due to the fuzziness of operation mechanism, the changes of environment, the complexity of conditions, the limitations of operating, and so on. One of the main tasks of the grey system theory is to find the mathematical relationship and the changing rules among the factors or within one factor based on the behavior feature data of social, economic, ecological system, and so on. Furthermore, grey system theory holds that any random process is a grey quantum that changes within certain ranges and time zones, so a random process is viewed as a grey process.

6.1.1 Grey Number, Whitening of Grey Number, and Grey Degree

Grey number is an expression of the grey system behavior character, and it is the basic "unit" or "cell" of the grey system. A grey number is such a number whose exact value is unknown but the range within which the value lies is known. In applications, a grey number is an uncertain number of an interval or a set of numbers.

A fundamental question of the grey system theory is how to whiten grey numbers and, in essence, how to find the largest element from a grey-element space (if it exists). The grey degree of grey numbers reflects the uncertainty degree of the grey system. The greater the uncertainty, the higher the grey degree; the smaller the uncertainty, the lower the grey degree. To keep the consistency among symbols of the book, the symbols involved are slightly different from the original literature, such as the symbol of the grey number $\otimes$ was revised to x.

6.1.1.1 Types of Grey Numbers

Grey numbers can be divided into the following types:

1. *Grey numbers with only lower limits*: The grey number with a lower limit but no upper limit is denoted as $x \in [\underline{a}, \infty]$, where $\underline{a}$ is the lower limit of the grey number x and it is a definite number.
2. *Grey numbers with only upper limits*: The grey number with only an upper limit is written as $x \in [-\infty, \overline{a}]$, where $\overline{a}$ is the upper limit of the grey number x and it is a definite number.
3. *Interval grey numbers*: A grey number with both a lower limit $\underline{a}$ and an upper limit $\overline{a}$ is called an interval grey number, denoted as $x \in [\underline{a}, \overline{a}]$.

4. *Continuous grey numbers and discrete grey numbers*: The grey numbers taking on a finite number of values or a countable number of values in an interval are called discrete grey numbers. And those continuously taking values that cover an interval are continuous grey numbers.
5. *Black and white numbers*: When $x \in [-\infty, +\infty]$, that is, when the lower limit and upper limit of the grey number are both infinite, the grey number is called black number; when $x \in [\underline{a}, \overline{a}]$ and $\underline{a} = \overline{a}$, the grey number is called white number.
6. *Essential grey numbers and nonessential grey numbers*: Essential grey numbers refer to the ones that cannot or temporarily cannot find a white number as its "representative" number. For example, the general value predicted in advance, the total energy of the universe, and the age that is accurate to the second or microsecond are all essential grey numbers. While nonessential grey numbers refer to the grey number called the whitening value, for which we can find a white number as its "representative" with the help of a priori information or some other means, the white number is called the whitening value of the corresponding grey number.

According to the nature of the grey, grey numbers can be divided into three types—grey numbers of information, conceptual grey numbers, and grey numbers of layer:

1. Grey numbers of information are those whose values cannot be certain due to temporary shortage of information. For example, the output of a certain area's summer grain is expected to be more than one million tons this year. This is a grey number of information. Once the predicted time comes, it will become a completely definite number.
2. Among all conceptual grey numbers, some are also called grey numbers of wish, which means that these grey numbers are formed based on people's wishes and thoughts. For example, a college assumes a national key scientific and technological subject, and hopes that scientific research funding is no less than 60 million yuan, and the more the better.
3. Grey numbers of layer are those formed by changing layers. For example, for the number of the name called Li Yang, there is only one in a certain university, while the city's universities have the numbers from 6 to 8, thus it is a grey number, if considered in the national level, and the number is more unclear.

6.1.1.2 Whitenization of Grey Numbers and Grey Degree

It is relatively easy to whiten a grey number that vibrates around a base value. And the base value can be denoted as the whiting value. For example, if some enterprise's output value is about 380 million Yuan in 2006, so the whiting value of the enterprise output value is 380 million Yuan. For a general interval grey number $x \in [a_1, a_2]$, we take $\tilde{x} \in \alpha a + (1 - a)b, \alpha \in [0, 1]$ as its whitening value.

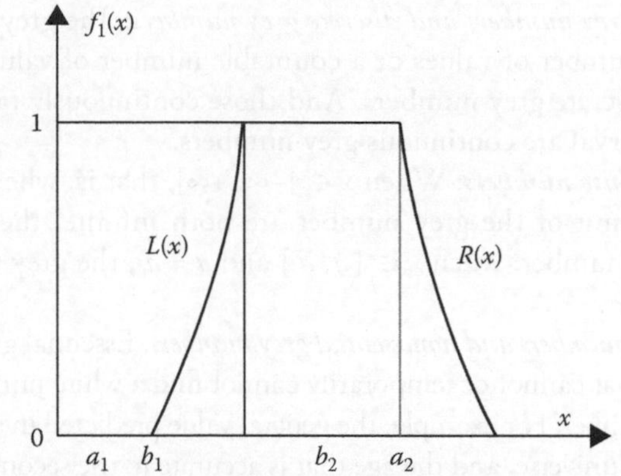

Figure 6.1 Typical weight functions of whitenization.

A grey number can usually be whitened through a whitening weight function, which is usually designed based on the given information, and there is no fixed pattern.

Typical weight functions of whitenization $f\,[a_1, b_1, b_2, a_2]$ are shown in Figure 6.1. Assumed that

$$f_1(x) = \begin{cases} L(x) & x \in [a_1, b_1) \\ 1 & x \in [b_1, b_2] \\ R(x) & x \in (b_2, a_2] \end{cases}$$

where

$L(x)$ is an increasing function of the left

$R(x)$ is a right down function

$[b_1, b_2]$ is the peak district

a_1 is the starting point

a_2 is the ending point

b_1, b_2 are turning points, the grey number $x \in [a_1, a_2]$

The continuous functions with fixed starting and ending points, and values increasing on the left and decreasing on the right are called typical weight functions of whitenization.

According to the grey system theory,

$$g^\circ = \frac{2\,|b_1 - b_2|}{b_1 + b_2} + \max\left\{\frac{|a_1 - b_1|}{b_1}, \frac{|a_2 - b_2|}{b_2}\right\} \tag{6.1}$$

g° is called the grey degree of the grey number x for the typical weight function of whitenization, as shown in Figure 6.1.

The expression (6.1) of the grey degree consists of two parts, in which the first one represents the impact of the size of the peak district on grey degree and the other represents the impact of the coverage of the $L(x)$ and $R(x)$ on grey degree. According to this definition, for the grey degree of a grey number, the greater the peak area and the larger the coverage area of $L(x)$ and $R(x)$, the greater the grey degree.

Based on the length of the information field of the grey number $l(x)$ and the mean-value whitenization (number) of the grey number $\hat{x}$, Professor Sifeng Liu put forward an axiomatic definition of the grey degree of the grey number in 1996:

$$g^{\circ} = \frac{l(x)}{\hat{x}} \tag{6.2}$$

However, for the two kinds of definition mentioned above, there are still some questions as below:

1. Not normative. For example, when the length of grey interval tends to be infinite, the degrees of greyness in the definitions of the formula (6.1) and formula (6.2) would possibly tend to infinity.
2. The grey degree of the grey number of the zero sum is not defined. For example, when $b_1 = b_2 = 0$ in the formula (6.1) and $\hat{x} = 0$ in the formula (6.2), there are no definitions of the corresponding grey degrees.

To cover these shortages, Professor Sifeng Liu put forward the following justice system for the definition of grey degree in 2004.

Definition 6.1 Assume that the background of introduction of a grey number, and let $m(x)$ be the measure of the field on which the grey number x is defined. Then, the grey degree g° of the grey number x satisfies the following axioms:

1. $0 \leq g^{\circ} \leq 1$.
2. When $x \in [a_1, a_2]$, $a_1 \leq a_2$, if $a_1 = a_2$, there is $g^{\circ} = 1$.
3. $g(U) = 1$.
4. g° is proportional to $m(x)$, and is in reverse proportion to $m(U)$.

To make grey degree more feasible in practical application, in 2005, Professor Sifeng Liu further put forward five levels of grey degree and constructed a new algorithm of interval grey number.

6.1.2 Grey Sequence Generation

Grey system finds changing rules of raw data by processing them, it is an approach to find the potential regulations of data through the data itself, and it is called grey sequence generating. According to grey system theory, despite the apparent complexity and destabilized data of the objective system, it always has a combined function, so there is some sort of internal rules inevitability, and the key point is to choose an appropriate way to explore and use it. For any grey sequence, we can weaken its randomness to give prominence to its regularity by a certain kind of generating operation. For example, for the original data sequence $X^{(0)} = (1, 2, 1, 5, 3)$, there is no obvious regularity, while it can be converted to monotone increasing sequence $X^{(1)} = (1, 3, 4, 5, 7, 5)$ with 1-AGO, as shown in Figure 6.2.

Compare the original sequence $X^{(0)}$ with grey generated sequence $X^{(1)}$ in the Figure 6.2, we can see that the curve of $X^{(0)}$ has much more random fluctuations and larger change range, while $X^{(1)}$ with 1-AGO has shown clear growth regularity.

Grey generating possesses the following significances:

1. To unify the goal nature (polarity) of sequences to provide polarity samples uniformly for grey decision-making.
2. To transfer the undulating sequences into monotone raising ones, so it is good for grey modeling.
3. To reveal the raising tendency hided in grey sequences, so the detached sequences can be inverted into comparable ones.

General quasi-smooth nonnegative sequence will reduce randomness and take on the similarity to the regularity of exponential growth after AGO. The smoother the original sequence, the more obvious the index rules after generating.

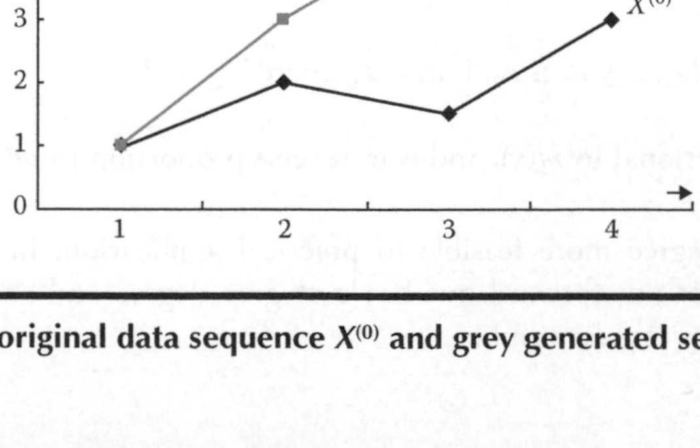

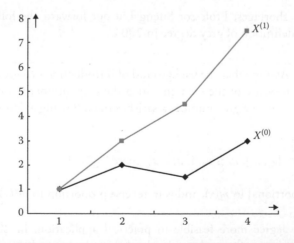

Figure 6.2 An original data sequence $X^{(0)}$ and grey generated sequence.

Definition 6.2 Let $X = (x(1), x(2), \ldots, x(n))$ be a sequence, then

$$\sigma(k) = \frac{x(k)}{x(k-1)}; \quad k = 2,3,\ldots,n \tag{6.3}$$

is called stepwise ratios of the sequence, while

$$\rho(k) = \frac{x(k)}{\sum_{i=1}^{k-1} x(i)}; \quad k = 2,3,\ldots,n \tag{6.4}$$

is called smooth ratios of the sequence. There are the following relations existing between stepwise ratios and smooth ratios:

$$\sigma(k+1) = \frac{\rho(k+1)}{\rho(k)}(1+\rho(k)); \quad k = 2,3,\ldots,n$$

Definition 6.3 If the sequence X satisfies the following conditions

1. $(\rho(k+1)/\rho(k)) < 1$; $k = 2, 3, \ldots, n-1$
2. $\rho(k) \in [0, \varepsilon]$; $k = 3, 4, \ldots, n$
3. $\varepsilon < 0.5$

then X is called quasi-smooth sequence.

Definition 6.4 Assume that sequence $X = (x(1), x(2), \ldots, x(n))$, if

1. For $\forall k$, $\sigma(k) \in (0, 1)$, then the sequence X has a negative grey exponential rule.
2. For $\forall k$, $\sigma(k) \in (1, b)$, then the sequence X has a positive grey exponential rule.
3. For $\forall k$, $\sigma(k) \in [a, b]$, $b - a = \delta$, then the sequence X has a grey exponential rule whose absolute grey degree is δ.
4. $\delta < 0.5$, then the sequence X has a quasi-exponential law.

If $X(0)$ is a quasi-smooth nonnegative sequence, then the sequence $X(1)$ that is from $X(0)$ with 1-AGO has a quasi-exponential rule, and this is the theoretical basis of the grey system modeling. Because economic systems, ecosystems, and agricultural systems can all be regarded as generalized energy systems, while the accumulation and release of energy has exponential rules generally, exponential model of grey system theory has a wide range of adaptability.

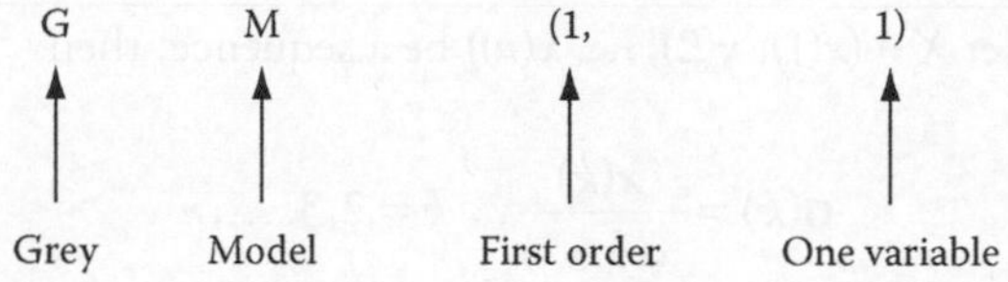

Figure 6.3 The symbol GM(1, 1).

6.1.3 GM(1, 1) Model

Definition 6.5 Assume that $X^{(0)} = (x^{(0)}(1), x^{(0)}(2), \ldots, x^{(0)}(n))$, $X^{(1)} = (x^{(1)}(1), x^{(1)}(2), \ldots, x^{(1)}(n))$, the equation

$$x^{(0)}(k) + ax^{(1)}(k) = b \tag{6.5}$$

is called the original GM(1, 1) model (the original form of GM(1, 1) model).

The meaning of the symbol GM(1, 1) is shown in Figure 6.3.

Definition 6.6 Assume that $X^{(0)}$, $X^{(1)}$ is defined as shown in the Definition 6.1,

$$Z^{(1)} = (z^{(1)}(2), z^{(1)}(3), \ldots, z^{(1)}(n))$$

where, $z^{(1)}(k) = (1/2)(x^{(1)}(k) + x^{(1)}(k-1))$, the equation

$$x^{(0)}(k) + az^{(1)}(k) = b \tag{6.6}$$

is called the basic GM(1, 1) (the basic form of GM(1, 1) model).

Proposition 6.1 Assume that $X^{(0)}$ is a nonnegative sequence:

$$X^{(0)} = (x^{(0)}(1), x^{(0)}(2), \ldots, x^{(0)}(n)), \quad \text{where } x^{(0)}(k) \geq 0,\ k = 1, 2, \ldots, n;$$

$X^{(1)}$ is the 1-AGO sequence of $X^{(0)}$:

$$X^{(1)} = (x^{(1)}(1), x^{(1)}(2), \ldots, x^{(1)}(n)), \quad \text{where } x^{(1)}(k) = \sum_{i=1}^{k} x^{(0)}(i),\ k = 1, 2, \ldots, n;$$

$Z^{(1)}$ is the mean sequences of $X^{(1)}$ given by

$$Z^{(1)} = (z^{(1)}(2), z^{(1)}(3), \ldots, z^{(1)}(n)), \quad \text{where } z^{(1)}(k) = (1/2)(x^{(1)}(k) + x^{(1)}(k-1)),\ k = 2, 3, \ldots, n.$$

If $\hat{a} = (a, b)^T$ is a sequence of parameters and

$$Y = \begin{bmatrix} x^{(0)}(2) \\ x^{(0)}(3) \\ \vdots \\ x^{(0)}(n) \end{bmatrix}, \quad B = \begin{bmatrix} -z^{(1)}(2) & 1 \\ -z^{(1)}(3) & 1 \\ \vdots & \vdots \\ -z^{(1)}(n) & 1 \end{bmatrix} \tag{6.7}$$

then the least-squares estimate parameter sequence of this grey differential equation $x^{(0)}(k) + az^{(1)}(k) = b$ satisfies $\hat{a} = (B^T B)^{-1} B^T Y$.

Definition 6.7 Assume that $X^{(0)}$ is a nonnegative sequence, $X^{(1)}$ is the 1-AGO sequence of $X^{(0)}$, and $Z^{(1)}$ is the mean sequences of $X^{(1)}$. If $[a, b]^T = (B^T B)^{-1} B^T Y$, the equation $dx^{(1)}/dt + ax^{(1)} = b$ is called the whitening equation of the grey differential equation $x^{(0)}(k) + az^{(1)}(k) = b$, and is also known as the shadow of equation.

Proposition 6.2 Assume that B, Y, $\hat{a}$ are the same as in Proposition 6.5, if $\hat{a} = [a, b]^T = (B^T B)^{-1} B^T Y$, then the following holds true:

1. The solution of the whitenization function $dx^{(1)}/dt + ax^{(1)} = b$ is given by

$$x^{(1)}(t) = \left(x^{(1)}(0) - \frac{b}{a} \right) e^{-at} + \frac{b}{a} \tag{6.8}$$

2. The time response sequence of the GM(1, 1) grey differential equation $x^{(0)}(k) + az^{(1)}(k) = b$ is given by

$$\hat{x}^{(1)}(k+1) = \left(x^{(1)}(0) - \frac{b}{a} \right) e^{-ak} + \frac{b}{a}; \quad k = 1, 2, \ldots, n \tag{6.9}$$

3. Let $x^{(1)}(0) = x^{(0)}(1)$, then

$$\hat{x}^{(1)}(k+1) = \left(x^{(0)}(1) - \frac{b}{a} \right) e^{-ak} + \frac{b}{a}; \quad k = 1, 2, \ldots, n \tag{6.10}$$

4. Restore the value

$$\hat{x}^{(0)}(k+1) = \alpha^{(1)}\hat{x}^{(1)}(k+1) = \hat{x}^{(1)}(k+1) - \hat{x}^{(1)}(k); \quad k = 1, 2, \ldots, n \tag{6.11}$$

The parameters a and b in GM(1, 1) model are called development coefficient and grey action quantity, respectively.

Example 6.1 Assume that the original sequence $X^{(0)} = (x^{(0)}(1), x^{(0)}(2), x^{(0)}(3), x^{(0)}(4), x^{(0)}(5)) = (3.167, 6.828, 10.984, 15.308, 20.27)$. Try to simulate $X^{(0)}$ with GM(1, 1) model, and calculate its simulation accuracy.

Solution

1. With 1-AGO for $X^{(0)}$, we can obtain $X^{(1)} = (x^{(1)}(1), x^{(1)}(2), x^{(1)}(3), x^{(1)}(4), x^{(1)}(5)) = (3.167, 6.828, 10.984, 15.308, 20.27)$.
2. Make the quasi-smooth test for $X^{(0)}$. According to $\rho(k) = x^{(0)}(k)/x^{(1)}(k-1)$, we can obtain $\rho(2) \approx 1.156$, $\rho(3) \approx 0.609$, $\rho(4) \approx 0.394 < 0.5$, $\rho(5) \approx 0.324 < 0.5$. So, when $k > 3$, the quasi-smooth condition is satisfied.
3. Test whether $X^{(1)}$ has the quasi-exponential law. According to $\sigma^{(1)}(k) = x^{(1)}(k)/x^{(1)}(k-1)$, we can obtain $\sigma^{(1)}(2) \approx 2.16$, $\sigma^{(1)}(3) \approx 1.61$, $\sigma^{(1)}(4) \approx 1.39$, $\sigma^{(1)}(5) = 1.32$

   If $k > 3$, $\sigma^{(1)}(k) \in [1, 1.5]$, and $\delta = 0.5$ satisfies the quasi-exponential law, then we can build GM(1,1) model for $X^{(1)}$.
4. Make the mean generated with consecutive neighbors of $X^{(1)}$. Assume that $Z^{(1)}(k) = 0.5x^{(1)}(k) + 0.5x^{(1)}(k-1)$, we can get $Z^{(1)} = (z^{(1)}(2), z^{(1)}(3), z^{(1)}(4), z^{(1)}(5)) = (4.9975, 8.906, 13.146, 17.789)$, thus,

$$B = \begin{bmatrix} -Z^{(1)}(2) & 1 \\ -Z^{(1)}(3) & 1 \\ -Z^{(1)}(4) & 1 \\ -Z^{(1)}(5) & 1 \end{bmatrix} = \begin{bmatrix} -4.9975 & 1 \\ -8.906 & 1 \\ -13.146 & 1 \\ -17.789 & 1 \end{bmatrix}, \quad Y = \begin{bmatrix} x^{(0)}(2) \\ x^{(0)}(3) \\ x^{(0)}(4) \\ x^{(0)}(5) \end{bmatrix} = \begin{bmatrix} 3.661 \\ 4.156 \\ 4.324 \\ 4.964 \end{bmatrix}$$

5. Make least-squares estimation for a sequence of parameters $\hat{a} = [a, b]^T$. We can get

$$\hat{a} = (B^T B)^{-1} B^T Y = \begin{bmatrix} -0.0957 \\ 3.2029 \end{bmatrix}$$

6. Determine the model

   $(dx^{(1)}/dt) - 0.0957x^{(1)} = 3.2029$, and the time response function $\hat{x}^{(1)}(k) = (x^{(0)}(1) - (b/a))e^{-a(k-1)} + b/a = 33.2884e^{0.09571}k - 33.4649$
7. Calculate the simulation value of $X^{(1)}$:

$$\hat{X}^{(1)} = (\hat{x}^{(1)}(1), \hat{x}^{(1)}(2), \hat{x}^{(1)}(3), \hat{x}^{(1)}(4), \hat{x}^{(1)}(5))$$
$$= (6.8463, 10.8951, 15.3505, 20.2535)$$

8. Restore to obtain the simulation value of $X^{(0)}$. By $\hat{x}^{(0)}(k) = \alpha^{(1)}\hat{x}^{(1)}(k) = \hat{x}^{(1)}(k) - \hat{x}^{(1)}(k-1)$, we can get

$$\hat{X}^{(0)} = (\hat{x}^{(0)}(1), \hat{x}^{(0)}(2), \hat{x}^{(0)}(3), \hat{x}^{(0)}(4), \hat{x}^{(0)}(5))$$
$$= (3.6793, 4.0488, 4.4555, 4.9030)$$

Table 6.1 Error Test Table

Sequence Number	*Actual Data* $x^{(0)}(k)$	*Analog Data* $\hat{x}^{(0)}(k)$	*Residual* $\varepsilon(k) = x^{(0)}(k) - \hat{x}^{(0)}(k)$	*Relative Error* $\Delta_k = \frac{\lvert \varepsilon(k) \rvert}{x^{(0)}(k)}$ *(Percent)*
2	3.278	3.230	−0.01826	0.50
3	3.337	3.3545	0.1072	2.58
4	3.390	3.4817	−0.1315	3.04
5	3.679	3.6136	0.05904	1.19

9. Test error. The results are shown in Table 6.1.
   Based on the data in Table 6.1, we can get the sum of residual squares

$$s = \varepsilon^T \varepsilon = [\varepsilon(2), \varepsilon(3), \varepsilon(4), \varepsilon(5)] \begin{bmatrix} \varepsilon(2) \\ \varepsilon(3) \\ \varepsilon(4) \\ \varepsilon(5) \end{bmatrix} = 0.03259$$

and the average relative error $\Delta = (1/4)\sum_{k=2}^{5} \Delta_k = 1.8271$ percent. From the sum of residual squares and the average relative error, we can see that the accuracy of the model is relatively high.

In addition to the index model in the GM(1, 1), there are a variety of forms of model and the improved forms of the model, such as differential model, the metabolic model, grey Verhulst model, and so on, which are listed in references at the end of the book, and we can choose suitable models according to actual problems.

6.1.4 Grey Correlation Analysis

Many methods in statistics, such as regression analysis, variance analysis, and principal component analysis, are all commonly used in the analysis of systems. However, these methods have the following pitfalls: it is required that all samples or populations satisfy certain typical probability distribution, and there is a linear relationship between factor data and system characteristics of the data, but the various factors are independent of each other. However, the grey correlation analysis can make up the lacks caused by mathematical statistics methods.

Less data and uncertainties generally present as limited sequences. In the grey system theory, each sequences represents as a factor, a model, a program, an act, and so on. To understand the borders of grey system and make analysis of the primary and secondary factors, the recognition mode, the optimization program, disposal behavior, and so on, it is necessary to make modeling analysis on the relationship between sequences, and it is called the grey correlation analysis.

The basic idea of grey correlation analysis is to see whether the relation is close or not by the similarity among the geometrical shapes of sequence curves. The closer the sequence curves are, the greater the correlation is. And vice versa.

To make analysis of an abstract system or phenomenon, first of all we have to select the data sequence that reflects the characteristics of system behavior, thus this is called to find the volume mapping of system behavior and indirectly characterize with volume mapping system behavior. Such as the national average number of years of education to reflect the level of education, with the incidence of criminal cases to reflect the face of social order and public order, and the registered number of hospitals to reflect the level of national health. With the characteristics of system behavior data and data relevant factors, we can make all the graphics sequence and analysis according to the visual. For example, total agricultural output value in a certain area X_0, output value of farming X_1, livestock output X_2, and fruit industry in GDP X_3, the total of six years statistics from 1997 to 2002 are as follows:

$$X_0 = (18, 20, 22, 35, 41, 46), \quad X_1 = (8,11,12,17,24,29),$$

$$X_2 = (3,2,7,4,11,6), \quad X_3 = (5,7,7,11,5,10)$$

The curves of each data sequence $X_i (i = 0, 1, 2, 3)$ are shown in Figure 6.4.

As can be seen from Figure 6.4, agricultural output value curve is most similar to farming output curve, while animal husbandry output value curve and fruit

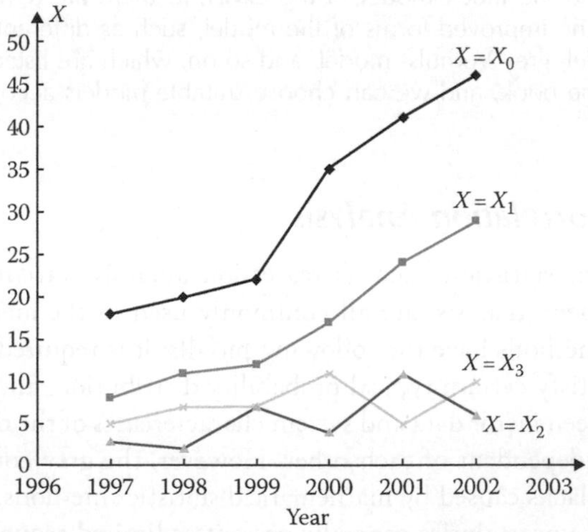

Figure 6.4 Curves of sequence $X_i (i = 0, 1, 2, 3)$.

production curve are larger from and agricultural output value curve in the geometric shape. In accordance with the needs of practical problems, we can do further research and quantity analysis.

In fact, grey correlation analysis may be seen as a reference system to compare with a whole system. It is well known to us, distance space is a comparison method of the number of measurement, however, there is a lack of integrity due to the comparison between only two objects. Point-set topology characterizes the neighborhood that the overall does not have digital measurement. Through combination of distance space and point-set topology, we can get grey correlation space, and it can be expressed specific as follows:

$$\text{Distance space} + \text{Point-set topology} = \text{Grey correlation space}$$

Grey correlation analysis has the following characteristics: less data (each sequence can be as little as three data), it does not consider the distribution of data, and the calculation is simple.

The difference of the grey correlation analysis and traditional mathematical analysis is that grey correlation analysis provides a sequence analysis or summary of the framework of system behavior, even if in the case that there is little information, it can also be completed. Grey correlation analysis has been successfully applied to optimization of urban roads rehabilitation programs, petroleum, mineral prospecting, medicine, and other fields. Grey correlation analysis is necessary for analysis of these relationships among the objects that are inherent unclear with behavior mechanism, or objects with scarce behavioral data, or problems lack of experience to deal with.

Proposition 6.3 Assume that $m + 1$ behavioral sequences of a system are given as follows:

$$\begin{cases} X_0 = (x_0(1), x_0(2), \ldots, x_0(n)) \\ X_1 = (x_1(1), x_1(2), \ldots, x_1(n)) \\ \quad\cdots \\ X_i = (x_i(1), x_i(2), \ldots, x_i(n)) \\ \quad\cdots \\ X_m = (x_m(1), x_m(2), \ldots, x_m(n)) \end{cases}$$

Let

$$\gamma(x_0(k), x_i(k)) = \frac{\min_i \min_k |x_0(k) - x_i(k)| + \xi \max_i \max_k |x_0(k) - x_i(k)}{|x_0(k) - x_i(k)| + \xi \max_i \max_k |x_0(k) - x_i(k)|}$$

and

$$\gamma(X_0, X_i) = \frac{1}{n}\sum_{k=1}^{n} \gamma(x_0(k), x_i(k))$$

where $\xi \in (0, 1)$ is called distinguishing coefficient, which we generally choose 0.5 according to the least information principle; $\gamma(X_0, X_i)$ is called the degree of grey correlation of X_i with respect to X_0, and $\gamma(x_0(k), x_i(k))$ the correlation coefficient of X_i with respect to X_0 at point k.

$\gamma(X_0, X_i)$ satisfies the following four axioms for grey correlation:

1. The property of normality

$$0 \le \gamma(x_0(k), x_i(k)) \le 1, \quad X_0 = X_i \Rightarrow \gamma(x_0(k), x_i(k)) = 1$$

2. The property of wholeness
   For any $X_i, X_j \in X = \{X_s \mid s = 0, 1, 2, \ldots, m, m > 2\}$, we have

$$\gamma(X_i, X_j) \ne \gamma(X_j, X_i), \quad i \ne j$$

3. The property of pair symmetry
   For $X_i, X_j \in X$, we have

$$\gamma(X_i, X_j) = \gamma(X_j, X_i) \Leftrightarrow X = \{X_i = X_j\}$$

4. The property of closeness

The smaller the absolute value $|x_0(k) - x_i(k)|$ is, the bigger $\gamma(x_0(k), x_i(k))$ is. The computing steps of grey degree of grey correlation are as follows:

Step 1: Calculate the initial value image (or mean-value image) of different sequences. Let

$$X_i' = X_i / x_i(1) = (x_i'(1), x_i'(2), \ldots, x_i'(n)), \quad i = 0, 1, 2, \ldots, m$$

Step 2: Calculate minus sequences. Let

$$\Delta_i(k) = |\, x_0'(k) - x_i'(k) \,|, \quad \text{where } i = 0, 1, 2, \ldots, m,\ k = 1, 2, \ldots, n$$

Step 3: Calculate the maximum minus value and the minimum minus value. Let

$$M = \max_i \max_k \Delta_i(k), \quad m = \min_i \min_k \Delta_i(k)$$

Step 4: Calculate correlation coefficient. Let

$$\gamma_{0i}(k) = \frac{m + \xi M}{\Delta_i(k) + \xi M}, \quad \xi \in (0,1), i = 0,1,2,\ldots,m, k = 1,2,\ldots,n$$

Step 5: Calculate degrees of grey correlation. Let

$$\gamma_{0i}(k) = \frac{1}{n}\sum_{k=1}^{n} \gamma_{0i}(k)$$

Example 6.2 In a study area, some behavioral data for industry, agriculture, transportation, and business are provided as follows:

Industry: $X_1 = (x_1(1), x_1(2), x_1(3), x_1(4)) = (45.8, 43.4, 42.3, 41.9)$
Agriculture: $X_2 = (x_2(1), x_2(2), x_2(3), x_2(4)) = (39.1, 41.6, 43.9, 44.9)$
Transportation: $X_3 = (x_3(1), x_3(2), x_3(3), x_3(4)) = (3.4, 3.3, 3.5, 3.5)$
Business: $X_4 = (x_4(1), x_4(2), x_4(3), x_4(4)) = (6.7, 6.8, 5.4, 4.7)$

Compute the degree of grey correlation by using X_1 as a system's characteristic sequence.

Solution

1. Calculate initial values images
   According to $X_i' = X_i / x_i(1) = (x_i'(1), x_i'(2), \ldots, x_i'(n))$, $i = 1, 2, 3, 4$, we can obtain

$$X_1' = (1, 0.9475, 0.9235, 0.9138), \quad X_2' = (1, 1.063, 1.1227, 1.1483)$$

$$X_3' = (1, 0.97, 1.0294, 1.0294), \quad X_4' = (1, 1.0149, 0.805, 0.7015)$$

2. Calculate difference sequences
   According to $\Delta_i(k) = |x_0'(k) - x_i'(k)|$, $i = 2, 3, 4$, we can obtain

$$\Delta_2 = (0, 0.1155, 0.1992, 0.2335)$$

$$\Delta_3 = (0, 0.0225, 0.1059, 0.1146)$$

$$\Delta_4 = (0, 0.0674, 0.1185, 0.2123)$$

3. Calculate bipolar differences

$$M = \max_i \max_k \Delta_i(k) = 0.2335; \quad m = \min_i \min_k \Delta_i(k) = 0$$

4. To calculate incidence coefficients
   We take $\xi = 0.5$, thus we can determine $\gamma_{1i}(k) = 0.11675/(\Delta_i(k) + 0.11675)$; $i = 2, 3, 4$; $k = 1, 2, 3, 4$; and obtain

$$\gamma_{12}(1) = 1, \quad \gamma_{12}(2) = 0.503, \quad \gamma_{12}(3) = 0.3695, \quad \gamma_{12}(4) = 0.3333$$

$$\gamma_{13}(1) = 1, \quad \gamma_{13}(2) = 0.8384, \quad \gamma_{13}(3) = 0.5244, \quad \gamma_{13}(4) = 0.504$$

$$\gamma_{14}(1) = 1, \quad \gamma_{14}(2) = 0.634, \quad \gamma_{14}(3) = 0.4963, \quad \gamma_{14}(4) = 0.352$$

5. To calculate degree of grey correlation

$$\gamma_{12} = \frac{1}{4}\sum_{k=1}^{4} \gamma_{12}(k) = 0.551$$

$$\gamma_{13} = \frac{1}{4}\sum_{k=1}^{4} \gamma_{13}(k) = 0.717$$

$$\gamma_{14} = \frac{1}{4}\sum_{k=1}^{4} \gamma_{14}(k) = 0.621$$

According to the above results, we can know that industry is most closely associated with transport industry, secondly with commercial, and least with agriculture in the region.

Proposition 6.4 Assume that $X_i = (x_i(1), x_i(2), \ldots, x_i(n))$ stands for a system's behavioral sequence, and two sequences X_0 and X_i are of the same length and $s_i = \int_1^n (X_i - x_i(1))\,dt$, then

$$\varepsilon_{0i} = \frac{1 + |s_0| + |s_i|}{1 + |s_0| + |s_i| + |s_i - s_0|} \tag{6.12}$$

is called the absolute degree of grey correlation of X_0 and X_i.

Absolute degree of grey correlation ε_{0i} has the following properties:

1. $0 < \varepsilon_{0i} < 1$.
2. ε_{0i} has something to do with the geometry of the X_0 and X_i, but it has nothing to do with their relative position in space, that is to say, horizontal motion does not change the value of the absolute degree of correlation.

3. For any two sequences, there are something to do with, that is, ε_{0i} does not constantly equal to zero.
4. The greater degree of similarity in the geometry of X_0 and X_i, the larger $\varepsilon_0 i$.
5. X_0 is parallel to X_i, or X_i^0 swings around X_0^0, and when the area of X_i^0 above the X_0^0 equals the area below X_0^0, $\varepsilon_{0i} = 1$.
6. When any observational data of X_0 and X_i changes, ε_{0i} will change correspondingly.
7. When the length of X_0 and X_i changes, ε_{0i} will have a corresponding change.
8. $\varepsilon_{00} = \varepsilon_{ii} = 1$.
9. $\varepsilon_{0i} = \varepsilon_{i0}$.

Compared with the degree of grey correlation, the absolute degree of grey correlation satisfies the property of normality, pair symmetry, and closeness, but does not satisfy the property of wholeness.

Example 6.3 Assume the sequence

$$X_0 = (x_0(1), x_0(2), x_0(3), x_0(4), x_0(5), x_0(6), x_0(7)) = (10,9,15,14,15,16)$$

$$X_1 = (x_1(1), x_1(2), x_1(3), x_1(4), x_1(5), x_1(6), x_1(7)) = (46,58,70,77,84,91,98).$$

Please compute its absolute degree of grey correlation ε_{01}.

Solution

1. To make the image of zero starting point, we can calculate

$$X_0 = (x_0(1), x_0(2), x_0(3), x_0(4), x_0(5), x_0(6), x_0(7)) = (0,-1,5,4,4,5,6)$$

$$X_1 = (x_1(1), x_1(2), x_1(3), x_1(4), x_1(5), x_1(6), x_1(7)) = (0,12,24,31,38,45,52)$$

2. To calculate $|s_0|$, $|s_1|$, $|s_1 - s_0|$, we can obtain

$$|s_0| = \left|\sum_{k=2}^{6} x_0^0(k) + \frac{1}{2}x_0^0(7)\right| = 20, \quad |s_1| = \left|\sum_{k=2}^{6} x_1^0(k) + \frac{1}{2}x_1^0(7)\right| = 176$$

$$|s_1 - s_0| = \left|\sum_{k=2}^{6} x_1^0(k) - x_0^0(k)) + \frac{1}{2}(x_1^0(7) - x_0^0(7))\right| = 156$$

3. To compute absolute degree of grey correlation

$$\varepsilon_{01} = \frac{1+|s_0|+|s_1|}{1+|s_0|+|s_1|+|s_1-s_0|} = \frac{197}{353} = 0.5581$$

The fundamental idea of grey correlation analysis is that the closeness of a relationship is judged based on the similarity level of the geometric patterns of sequence curves, in view of the method that there are no special requirements for the size of the sample size and analysis does not require the typical distribution, thus it has broad applicability.

6.1.5 Grey Correlation Order

When we make system analysis and study the relationship between the behavior of the system characteristics and related factors, the main concern is usually the order of the size of degree, the system characteristic behavior sequence, and all relevant factors, not entirely the size of the correlation degree of the numerical.

Definition 6.8 Assume that $Y_1, Y_2, \ldots, Y_s$ are sequences of system's characteristic behaviors, and $X_1, X_2, \ldots, X_m$ are called behavioral sequences of relevant factors. If the sequences $Y_1, Y_2, \ldots, Y_s$ and $X_1, X_2, \ldots, X_m$ have the same length, and $r_{ij} (i = 1, 2, \ldots, s; j = 1, 2, \ldots, m)$ is the degree of grey correlation of Y_i and X_j, then

$$\Gamma = (r_{ij}) = \begin{bmatrix} r_{11} & r_{12} & \cdots & r_{1m} \\ r_{21} & r_{22} & \cdots & r_{2m} \\ \cdots & \cdots & \cdots & \cdots \\ r_{s1} & r_{s2} & \cdots & r_{sm} \end{bmatrix}$$

is called the matrix of grey correlation.

Similarly, the matrix of grey absolute correlation can be defined as

$$A = (\varepsilon_{ij}) = \begin{bmatrix} \varepsilon_{11} & \varepsilon_{12} & \cdots & \varepsilon_{1m} \\ \varepsilon_{21} & \varepsilon_{22} & \cdots & \varepsilon_{2m} \\ \cdots & \cdots & \cdots & \cdots \\ \varepsilon_{s1} & \varepsilon_{s2} & \cdots & \varepsilon_{sm} \end{bmatrix}$$

With the use of grey correlation matrix, we can make advantage analysis of the system characteristics or relevant factors.

Definition 6.9 Assume that $Y_i(i = 1, 2, \ldots, s)$ is sequence of system's characteristic behaviors, and $X_j(i = 1, 2, \ldots m)$ is called behavioral sequence of relevant factors, and

$$\Gamma = (r_{ij}) = \begin{bmatrix} r_{11} & r_{12} & \cdots & r_{1m} \\ r_{21} & r_{22} & \cdots & r_{2m} \\ \cdots & \cdots & \cdots & \cdots \\ r_{s1} & r_{s2} & \cdots & r_{sm} \end{bmatrix}$$

is the matrix of grey correlation, if there exists $k, i \in \{i = 1, 2, \ldots, s\}$ and it satisfies $r_{kj} \geq r_{ij}$ and $j = 1, 2, \ldots, m$, then we say that the system's characteristic Y_k is more favorable than the system's characteristic Y_i.

Definition 6.10 Assume that $Y_i(i = 1, 2, \ldots, s)$ is sequence of system's characteristic behaviors, and $X_j(i = 1, 2, \ldots, m)$ is called behavioral sequence of relevant factors, and

$$\Gamma = (r_{ij}) = \begin{bmatrix} r_{11} & r_{12} & \cdots & r_{1m} \\ r_{21} & r_{22} & \cdots & r_{2m} \\ \cdots & \cdots & \cdots & \cdots \\ r_{s1} & r_{s2} & \cdots & r_{sm} \end{bmatrix}$$

is the matrix of grey incidences, if there exists $k, i \in \{i = 1, 2, \ldots, s\}$ and it satisfies $r_{kl} \geq r_{ij}$ and $i = 1, 2, \ldots, s$, then the factor X_l is said to be more favorable that the factor X_j.

Example 6.4 Assume that the sequences of system's characteristic behaviors

$$Y_1 = (170, 174, 197, 216.4, 235.8), \quad Y_2 = (57.55, 70.74, 76.8, 80.7, 89.85)$$

$$Y_3 = (68.56, 70, 85.38, 99.83, 103.4)$$

and behavioral sequences of relevant factors

$$X_1 = (308.58, 310, 295, 346, 367), \quad X_2 = (195, 4, 189.9, 189.2, 205, 222.7)$$

$$X_3 = (24.6, 21, 12.2, 15.1, 14.57), \quad X_4 = (20, 25.6, 23.3, 29.2, 30)$$

$$X_5 = (18.98, 19, 22.3, 23.5, 27.655)$$

Try to make advantage analysis with the matrix of grey absolute correlation.

Solution: To calculate as the starting point of zero on every behavior sequence, we can get the following results:

$$Y_1^0 = (0,4,27,46.4,65.8),\quad Y_2^0 = (0,13.19,19.25,23.15,32.3)$$

$$Y_3^0 = (0,1.44,16.82,31.27,34.84),\quad X_1^0 = (0,1.42,-13.58,37.42,58.42)$$

$$X_2^0 = (0,-5.5,-8.2,9.6,27.3),\quad X_3^0 = (0,-3.6,-12.4,-9.5,-10.03)$$

$$X_4^0 = (0,5.6,3.3,9.2,10),\quad X_5^0 = (0,0.02,3.32,4.52,8.675)$$

Corresponding to the system characteristics Y_1, we can calculate

$$|Y_{s1}| = \left|\sum_{k=2}^{4} y_1^0(k) + \frac{1}{2} y_1^0(5)\right| = \left|4+27+46.4+\frac{1}{2}\times 65.8\right| = 110.3$$

$$|X_{s1}| = \left|\sum_{k=2}^{4} x_1^0(k) + \frac{1}{2} x_1^0(5)\right| = \left|1.42+(-13.58)+37.42+\frac{1}{2}\times 58.42\right| = 54.47$$

$$|X_{s1} - Y_{s1}| = \left|\sum_{k=2}^{4} (x_1^0(k) - y_1^0(k)) + \frac{1}{2}(x_1^0(5) - y_1^0(5))\right| = 55.9$$

$$\varepsilon_{11} = \frac{1+|Y_{s1}|+|X_{s1}|}{1+|Y_{s1}|+|X_{s1}|+|X_{s1}-Y_{s1}|} = \frac{1+110.3+54.47}{1+110.3+54.47+55.9} = 0.748$$

$$|X_{s2}| = \left|\sum_{k=2}^{4} x_2^0(k) + \frac{1}{2} x_2^0(5)\right| = \left|(-5.5)+(-8.2)+9.6+\frac{1}{2}\times 27.3\right| = 9.55$$

$$|X_{s2} - Y_{s1}| = \left|\sum_{k=2}^{4} (x_2^0(k) - y_1^0(k)) + \frac{1}{2}(x_2^0(5) - y_1^0(5))\right| = 100.75$$

$$\varepsilon_{12} = \frac{1+|Y_{s1}|+|X_{s2}|}{1+|Y_{s1}|+|X_{s2}|+|X_{s2}-Y_{s1}|} = \frac{1+110.3+9.55}{1+110.3+9.55+100.75} = 0.545$$

Similarly, we can obtain the following results:

$$\varepsilon_{13} = \frac{1+|Y_{s1}|+|X_{s3}|}{1+|Y_{s1}|+|X_{s3}|+|X_{s3}-Y_{s1}|} = 0.502$$

$$\varepsilon_{14} = \frac{1+|Y_{s1}|+|X_{s4}|}{1+|Y_{s1}|+|X_{s4}|+|X_{s4}-Y_{s1}|} = 0.606$$

$$\varepsilon_{15} = \frac{1+|Y_{s1}|+|X_{s5}|}{1+|Y_{s1}|+|X_{s5}|+|X_{s5}-Y_{s1}|} = 0.557$$

Corresponding to the system characteristics Y_2, we can similarly obtain

$$\varepsilon_{21} = 0.880, \quad \varepsilon_{22} = 0.570, \quad \varepsilon_{23} = 0.502, \quad \varepsilon_{24} = 0.663, \quad \varepsilon_{25} = 0.588$$

Similarly, for the system characteristics Y_3, we can obtain

$$\varepsilon_{31} = 0.907, \quad \varepsilon_{32} = 0.574, \quad \varepsilon_{33} = 0.503, \quad \varepsilon_{34} = 0.675, \quad \varepsilon_{35} = 0.594$$

Thus, we can obtain absolute incidence matrix A, where

$$A = (\varepsilon_{ij}) = \begin{bmatrix} \varepsilon_{11} & \varepsilon_{12} & \varepsilon_{13} & \varepsilon_{14} & \varepsilon_{15} \\ \varepsilon_{21} & \varepsilon_{22} & \varepsilon_{23} & \varepsilon_{24} & \varepsilon_{25} \\ \varepsilon_{31} & \varepsilon_{32} & \varepsilon_{33} & \varepsilon_{34} & \varepsilon_{35} \end{bmatrix}$$

$$= \begin{bmatrix} 0.748 & 0.545 & 0.502 & 0.606 & 0.557 \\ 0.880 & 0.570 & 0.502 & 0.663 & 0.588 \\ 0.907 & 0.574 & 0.503 & 0.675 & 0.594 \end{bmatrix}$$

As line elements in matrix A meet the $\varepsilon_{3j} > \varepsilon_{2j} > \varepsilon_{1j}$ while $j = 1, 2, 3, 4, 5$, according to the absolute correlation matrix A, so there exists $Y_3 > Y_2 > Y_1$, that is to say, characteristic of Y_3 is optimal, Y_2 second, Y_1 worst.

6.1.6 Grey Clustering Evaluation

6.1.6.1 Clusters of Grey Correlation

Definition 6.11 Assume that n observational objects and m characteristic data values for each of these objects have been collected. So, we have the sequences

$$\begin{cases} X_1 = (x_1(1), x_1(2), \ldots, x_1(n)) \\ X_2 = (x_2(1), x_2(2), \ldots, x_2(n)) \\ \qquad \cdots \\ X_m = (x_m(1), x_m(2), \ldots, x_m(n)) \end{cases}$$

For all $i \le j$, $i, j = 1, 2, \ldots, m$, we calculate the absolute degree ε_{ij} of correlation of X_i and X_j, and obtain the following upper triangular matrix A

$$A = \begin{bmatrix} \varepsilon_{11} & \varepsilon_{12} & \cdots & \varepsilon_{1m} \\ & \varepsilon_{22} & \cdots & \varepsilon_{2m} \\ & & \cdots & \cdots \\ & & & \varepsilon_{mn} \end{bmatrix}$$

where $\varepsilon_{ij} = 1$, $i = 1, 2, \ldots, m$. Then the previous matrix A is called the correlation matrix of the characteristic variables. If the given critical value is $r \in [0,1]$, which generally requests $r > 0.5$, and $\varepsilon_{ij} \geq r$ $(i \neq j)$, X_i and X_j is regarded as the same characteristic. The classification of characteristic variables $X_1, X_2, \ldots, X_m$ under a fixed critical value r is called a cluster of r grey correlation.

6.1.6.2 Cluster with Variable Weights

Definition 6.12 Assume that there exists n objects to be clustered according to m cluster criteria into s different grey classes. The clustering method based on the observational value of the ith object, $(i = 1, 2, \ldots, n)$, at the jth criterion, $(j = 1, 2, \ldots, m)$, in which the ith object is classified into the kth grey class $(k \in \{1, 2, \ldots, s\})$, is called a grey clustering. All the s grey class formed by the n objects, defined by their observational values at the criterion j, are called the j-criterion subclasses. The whitenization weight function of the kth subclass of the j-criterion is denoted as $f_j^k(\cdot)$.

Assume that the whitenization weight function $f_j^k(\cdot)$ of a j-criterion kth subclass is $f_j^k(\cdot)$ shown in Figure 6.5. Then the points $x_j^k(1), x_j^k(2), x_j^k(3), x_j^k(4)$ are called turning points of $f_j^k(\cdot)$. The typical whitenization functions are expressed as $f_j^k\left(x_j^k(1), x_j^k(2), x_j^k(3), x_j^k(4)\right)$, then it has

$$f_j^k(x) = \begin{cases} 0 & x \notin [x_j^k(1), x_j^k(4)] \\ \dfrac{x - x_j^k(1)}{x_j^k(2) - x_j^k(1)} & x \in [x_j^k(1), x_j^k(2)] \\ 1 & x \in [x_j^k(2), x_j^k(3)] \\ \dfrac{x_j^k(4) - x}{x_j^k(4) - x_j^k(3)} & x \in [x_j^k(3), x_j^k(4)] \end{cases} \tag{6.13}$$

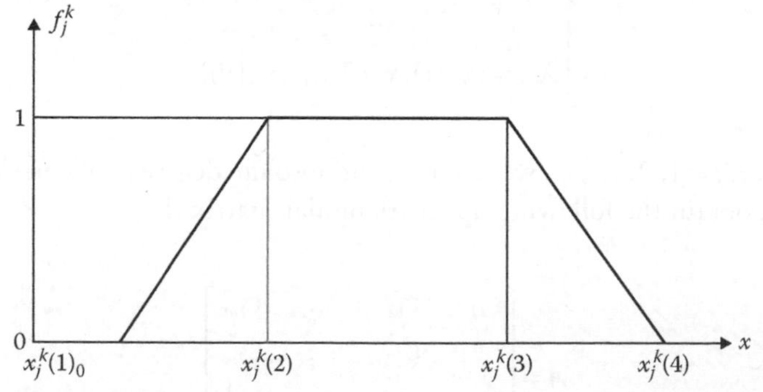

Figure 6.5 A typical whitenization function.

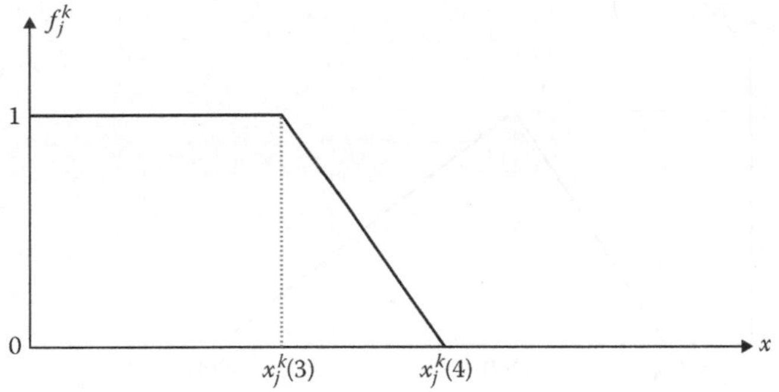

Figure 6.6 A whitenization function of lower measure.

If the whitenization weight function $f_j^k(\cdot)$ above does not have the first and the second turning points $x_j^k(1)$ and $x_j^k(2)$, as shown in Figure 6.6, then $f_j^k(\cdot)$ is called a whitenization weight function of lower measure, recorded as $f_j^k\left[-,-,x_j^k(3),x_j^k(4)\right]$, and it has

$$f_j^k(x)=\begin{cases} 0 & x\notin[0,x_j^k(4)] \\ 1 & x\in[0,x_j^k(3)] \\ \dfrac{x_j^k(4)-x}{x_j^k(4)-x_j^k(3)} & x\in[x_j^k(3),x_j^k(4)] \end{cases} \tag{6.14}$$

If the second point $x_j^k(2)$ and the third point $x_j^k(3)$ of the whitenization weight function $f_j^k(\cdot)$ coincide, as shown in Figure 6.7, then $f_j^k(\cdot)$ is called a whitenization function of moderate measure, recorded as $f_j^k\left[x_j^k(1),x_j^k(2),-,x_j^k(4)\right]$, and it has

$$f_j^k(x)=\begin{cases} 0 & x\notin[x_j^k(1),x_j^k(4)] \\ \dfrac{x-x_j^k(1)}{x_j^k(2)-x_j^k(1)} & x\in[x_j^k(1),x_j^k(2)] \\ \dfrac{x_j^k(4)-x}{x_j^k(4)-x_j^k(2)} & x\in[x_j^k(2),x_j^k(4)] \end{cases} \tag{6.15}$$

If the whitenization weight function $f_j^k(\cdot)$ does not have the third and fourth turning points, as shown in Figure 6.8, then $f_j^k(\cdot)$ is called a whitenization weight function of upper measure, recorded as $f_j^k\left[x_j^k(1),x_j^k(2),-,-\right]$, and it has

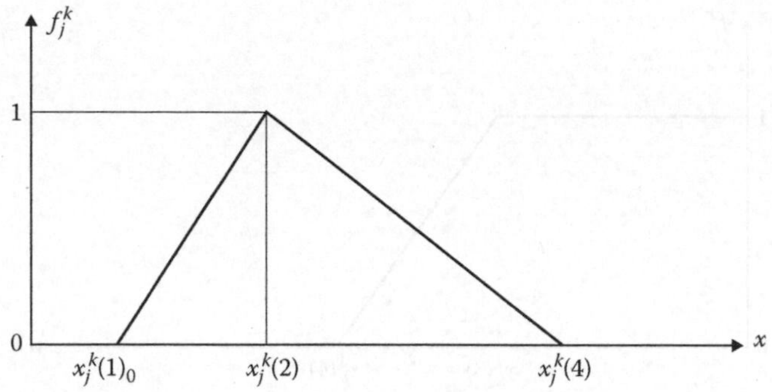

Figure 6.7 A whitenization function of middle measure.

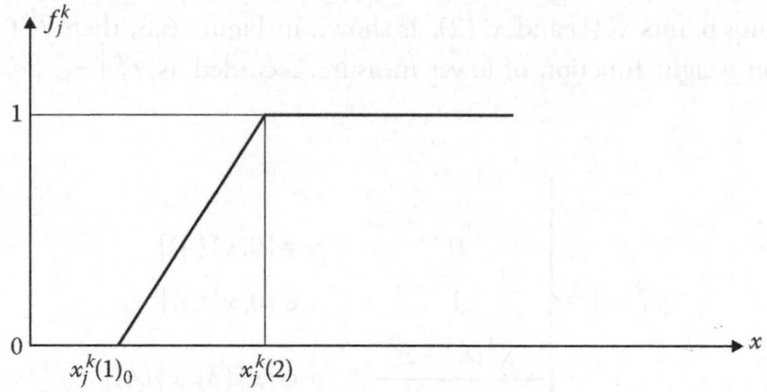

Figure 6.8 A whitenization function of upper measure.

$$f_j^k(x) = \begin{cases} 0 & x < x_j^k(1) \\ \dfrac{x - x_j^k(1)}{x_j^k(2) - x_j^k(1)} & x \in [x_j^k(1), x_j^k(2)] \\ 1 & x \geq x_j^k(2) \end{cases} \tag{6.16}$$

Definition 6.13 Assume that λ_j^k is the critical value for the kth subclass of the j-criterion. For a typical whitenization weight function, define $\lambda_j^k = (1/2)\left(x_j^k(2) + x_j^k(3)\right)$; for a whitenization weight function of lower measure, let $\lambda_j^k = x_j^k(3)$; and for a whitenization weight function of upper measure and the whitenization weight function of moderate measure, let $\lambda_j^k = x_j^k(2)$. Then

$$\eta_j^k = \frac{\lambda_j^k}{\sum_{j=1}^m \lambda_j^k} \tag{6.17}$$

is called the weight of the *j*-criterion with the *k*th subclass.

Definition 6.14 Assume that x_{ij} is the observational value of object *i* with criterion *j*, $f_j^k(\cdot)$ is a whitenization weight function of the *k*th subclass of the *j*-criterion, and η_j^k is the weight of the *j*-criterion with the *k*th subclass, then

$$\sigma_i^k = \sum_{j=1}^m f_j^k(x_{ij})\eta_j^k \tag{6.18}$$

is said to be the cluster coefficient for object *i* belonging to the *k*th grey class;

$$\sigma_i = (\sigma_i^1, \sigma_i^2, \ldots, \sigma_i^s) = \left(\sum_{j=1}^m f_j^1(x_{ij}) \cdot \eta_j^1, \sum_{j=1}^m f_j^2(x_{ij}) \cdot \eta_j^2, \ldots, \sum_{j=1}^m f_j^s(x_{ij}) \cdot \eta_j^s \right)$$

is called the cluster coefficient vector of object *i*; thus,

$$(\sigma_i^k) = \begin{bmatrix} \sigma_1^1 & \sigma_1^2 & \cdots & \sigma_1^s \\ \sigma_2^1 & \sigma_2^2 & \cdots & \sigma_2^s \\ \cdots & \cdots & \cdots & \cdots \\ \sigma_n^1 & \sigma_n^2 & \cdots & \sigma_n^s \end{bmatrix}$$

is called the cluster coefficient matrix. If $\max_{1 \le k \le s} \{\sigma_i^k\} = \sigma_i^{k^*}$, then we say that object *i* belongs to the grey class k^*.

Grey cluster with variable weights applies to the situation with the same meaning and dimensions of indicators. If there are more differences of significance and dimension and greater disparity of observations between different indicators in quantitative terms, grey cluster with variable weights should not be used.

Example 6.5 Assume that there are three economic zones and three grey cluster indicators that are "crop production income," "animal husbandry income," and "labor side occupation income." x_{ij} is the observational value of the ith economic zone with respect to the *j*th criterion ($j = 1, 2, \ldots, m$), and it is shown as the matrix *A*:

$$A = (x_{ij}) = \begin{bmatrix} x_{11} & x_{12} & x_{13} \\ x_{21} & x_{22} & x_{23} \\ x_{31} & x_{32} & x_{33} \end{bmatrix} = \begin{bmatrix} 80 & 20 & 100 \\ 40 & 30 & 30 \\ 10 & 90 & 60 \end{bmatrix}$$

Try to carry out the comprehensive cluster according to the high income, the medium income, and the low income.

Solution: Assume that the whitenization weight functions of these indicators are the following:

$$f_1^1[30,80,-,-],\quad f_1^2[10,40,-,70],\quad f_1^3[-,-,10,30]$$

$$f_2^1[30,90,-,-],\quad f_2^2[20,50,-,90],\quad f_2^3[-,-,20,40]$$

$$f_3^1[40,100,-,-],\quad f_3^2[30,60,-,90],\quad f_3^3[-,-,30,50]$$

That is,

$$f_1^1(x)=\begin{cases}0 & x<30\\ \dfrac{x-30}{80-30} & 30\le x<80,\\ 1 & x\ge 80\end{cases}\quad f_1^2(x)=\begin{cases}0 & x<10 \text{ or } x\ge 70\\ \dfrac{x-10}{40-10} & 10\le x<40\\ \dfrac{70-x}{70-40} & 40\le x<70\end{cases}$$

$$f_1^3(x)=\begin{cases}0 & x<0 \text{ or } x\ge 30\\ 1 & 0\le x<10\\ \dfrac{30-x}{30-10} & 10\le x<30\end{cases}$$

$$f_2^1(x)=\begin{cases}0 & x<30\\ \dfrac{x-30}{90-30} & 30\le x<90,\\ 1 & x\ge 90\end{cases}\quad f_2^2(x)=\begin{cases}0 & x<10 \text{ or } x\ge 90\\ \dfrac{x-20}{50-20} & 20\le x<50\\ \dfrac{90-x}{90-50} & 50\le x<90\end{cases}$$

$$f_2^3(x)=\begin{cases}0 & x<0 \text{ or } x\ge 40\\ 1 & 0\le x<20\\ \dfrac{40-x}{40-20} & 20\le x<40\end{cases}$$

$$f_3^1(x)=\begin{cases}0 & x<40\\ \dfrac{x-40}{100-40} & 40\le x<100,\\ 1 & x\ge 100\end{cases}\quad f_3^2(x)=\begin{cases}0 & x<30 \text{ or } x\ge 90\\ \dfrac{x-30}{50-30} & 30\le x<50\\ \dfrac{90-x}{90-50} & 50\le x<90\end{cases}$$

$$f_3^3(x)=\begin{cases}0 & x<0 \text{ or } x\ge 50\\ 1 & 0\le x<30\\ \dfrac{50-x}{50-30} & 30\le x<50\end{cases}$$

Then $\lambda_1^1=80,\ \lambda_2^1=90,\ \lambda_3^1=100;\ \lambda_1^2=40,\ \lambda_2^2=50,\ \lambda_3^2=60;\ \lambda_1^3=10,\ \lambda_2^3=20,$ $\lambda_3^3=30$, from $\eta_j^k=\lambda_j^k\Big/\sum_{j=1}^{3}\lambda_j^k$, we can obtain

$$\eta_1^1=\frac{80}{270},\quad \eta_2^1=\frac{90}{270},\quad \eta_3^1=\frac{100}{270},\quad \eta_1^2=\frac{40}{150},\quad \eta_2^2=\frac{50}{150},\quad \eta_3^2=\frac{60}{150}$$

$$\eta_1^3=\frac{10}{60},\quad \eta_2^3=\frac{20}{60},\quad \eta_3^3=\frac{30}{60}$$

When $i=1$, we can calculate

$$\begin{aligned}\sigma_i^k&=\sum_{j=1}^{m}f_j^k(x_{ij})\eta_j^k=\sigma_1^1=\sum_{j=1}^{3}f_j^1(x_{1j})\cdot\eta_j^1\\&=f_1^1(80)\times\left(\frac{80}{270}\right)+f_2^1(90)\times\left(\frac{90}{270}\right)+f_3^1(100)\times\left(\frac{100}{270}\right)\\&=0.6667\end{aligned}$$

Similarly, we can obtain $\sigma_1^1=0.6667,\ \sigma_1^2=0,\ \sigma_1^3=0.3333$, that is, $\sigma_1=\left(\sigma_1^1,\sigma_1^2,\sigma_1^3\right)=(0.6667,0,0.3333)$

If $i=2$, similarly, we can obtain $\sigma_2=\left(\sigma_2^1,\sigma_2^2,\sigma_2^3\right)=(0.0533,0.3778,0.6667)$
If $i=3$, similarly, we can obtain $\sigma_3=\left(\sigma_3^1,\sigma_3^2,\sigma_3^3\right)=(0.4568,0.3,0.1667)$

Thus, we can obtain the grey cluster coefficient matrix:

$$(\sigma_i^k)=\begin{bmatrix}\sigma_1^1 & \sigma_1^2 & \sigma_1^3\\ \sigma_2^1 & \sigma_2^2 & \sigma_2^3\\ \sigma_3^1 & \sigma_3^2 & \sigma_3^3\end{bmatrix}=\begin{bmatrix}0.6667 & 0 & 0.3333\\ 0.0533 & 0.3778 & 0.6667\\ 0.4568 & 0.3 & 0.1667\end{bmatrix}$$

6.1.6.3 Grey Cluster with Fixed Weights

When the clustering criterias have different meanings, dimensions, and sizes of observational data, applying variable weight clustering may lead to the problem that some criteria participate in the clustering process very weakly. There are two ways to solve this problem. The first is by transforming every indicator into a non-dimensional one with the usage of initial value or average operator, and then doing the cluster. This method deals with all cluster indicators with similar treatment, but it cannot reflect the difference made in the cluster process. The second is by defining a weight for each individual criterion before clustering, and the method of grey cluster with fixed weights belongs to this method.

Definition 6.15 Assume that x_{ij} $(i = 1,2,\ldots,n;\ j = 1,2,\ldots,m)$ is the observational value of object i with criterion j, and $f_j^k(\cdot)$ $(j = 1,2,\ldots,m;\ k = 1,2,\ldots,s)$ is the whitenization weight function of the kth subclass with the j-criterion.

1. If the weight η_j^k $(j = 1,2,\ldots,m;\ k = 1,2,\ldots,s)$ of the j-criterion has nothing to do with the kth subclass, that is, for any k_1, $k_2 \in \{1, 2,\ldots, s\}$, one always has that $\eta_j^{k_1} = \eta_j^{k_2}$, then the superscript η_j^k in the symbol will be omitted and written as $\eta_j (j = 1,2,\ldots,m)$ instead, and $\sigma_i^k = \sum_{j=1}^m f_j^k(x_{ij})\eta_j$ is called the fixed weight cluster coefficient for the object i belonging to the kth grey class.
2. If for any $(j = 1,2,\ldots,m)$, $\eta_j = (1/m)$ always holds true, then $\sigma_i^k = \sum_{j=1}^m f_j^k(x_{ij})\eta_j = (1/m)\sum_{j=1}^m f_j^k(x_{ij})$ is called equal weight cluster coefficient for the object i belonging to the kth grey class.

Classification according to the values of equal weight cluster coefficient is called equal weight grey cluster. Steps of equal weight grey cluster are the following:

Step 1: Give the whitenization weight function $f_j^k(\cdot)$ $(j = 1,2,\ldots,m;\ k = 1,2,\ldots,s)$ for the object i belonging to the kth grey class.

Step 2: Determine a cluster weight $\eta_j (j = 1,2,\ldots, m)$ to each criterion.

Step 3: According to the whitenization weight function $f_j^k(\cdot)$ $(j = 1,2,\ldots,m;\ k = 1,2,\ldots,s)$ of the Step 1 and Step 2 and a cluster weight $\eta_j (j = 1,2,\ldots,m)$ to each criterion, and $x_{ij} (i = 1, 2,\ldots, n;\ j = 1, 2,\ldots, m)$, calculate grey cluster coefficient with fixed weight $\sigma_i^k = \sum_{j=1}^m f_j^k(x_{ij})\eta_j$, $i = 1,2,\ldots,n;\ k = 1,2,\ldots,s$.

Step 4: If $\max_{1\le k\le s}\{\sigma_i^k\} = \sigma_i^{k^*}$, then judge that the object i belongs to the kth grey class.

6.2 Establishment of Decision Table Based on Grey Clustering

The decision table generated with the reduct by rough set approaches must be established in advance, but in many practical applications, we set up information systems only with datasets so that we are unable to obtain decision table, while grey clustering provides an approach by which the decision table is constructed from information system. And the steps are as follows:

1. Establish an information matrix with given attribute value.
2. Calculate cluster weight η_j^k (or given by experts) with the given whitenization weight function $f_j^k(\cdot)$.
3. From formula $\sigma_i^k = \sum_{j=1}^{m} f_j^k(x_{ij})\eta_j^k$, we can figure out cluster coefficient, and further determine object belonging to grey class.
4. Construct decision table with decision attribute value regarded as the object belonging to grey class.

Example 6.6 There are six scientific and technological personnel with intermediate technical titles in a company, an information system with quantitative evaluation indexes of ideological and moral, workload, operational level, scientific research and treatises, and foreign language proficiency is shown in Table 6.2 (where a_1 stands for ideological and moral, a_2 stands for workload, a_3 stands for operational level, a_4 stands for scientific research and treatises, and a_5 stands for foreign language proficiency). Establish the comprehensive evaluation decision table based on grey clustering.

Assume that the whitenization weight functions about five decision indicators with three grey classes are $f_j^1[70,90,-,-]$, $f_j^2[55,70,-,85]$, $f_j^3[-,-,40,65]$, respectively, that is,

Table 6.2 Comprehensive Evaluation Information System of Scientific and Technological Personnel

U	a_1	a_2	a_3	a_4	a_5
1	85	100	98	100	90
2	89	97	64	20	73
3	90	67	63	17	49
4	88	86	63	15	58
5	72	80	76	40	81
6	78	79	76	36	78

$$f_j^1(x)=\begin{cases}0 & x<70\\ \dfrac{x-70}{20} & 70\le x\le 90,\\ 1 & x>90\end{cases}\quad f_j^2(x)=\begin{cases}0 & x<55\\ \dfrac{x-55}{15} & 55\le x\le 70\\ \dfrac{85-x}{15} & 70<x\le 85\\ 0 & x>85\end{cases}$$

$$f_j^3(x)=\begin{cases}0 & x<0\\ 1 & 0\le x\le 40\\ \dfrac{65-x}{25} & 40<x\le 65\\ 0 & x>65\end{cases}$$

where $j = 1,2,\ldots,5$, meaning different indicator of the whitened weight function is identical.

Suppose that cluster weight (given by experts) of each indicator is, respectively,

$$\eta_1 = 0.22,\quad \eta_2 = 0.22,\quad \eta_3 = 0.22,\quad \eta_4 = 0.20,\quad \eta_5 = 0.14$$

Carry out cluster with the three grey classes according to competence, basic competence, and incompetence, and establish decision table (Table 6.3).

Solution: From formula $\sigma_i^k = \sum_{j=1}^{m} f_j^k(x_{ij})\eta_j^k$ $(i = 1,2,\ldots,6;\ k = 1,2,3)$, we can figure out cluster coefficient of each indicator and obtain the matrix of grey cluster coefficients:

Table 6.3 A Decision Table of the Comprehensive Evaluation of Scientific and Technological Personnel

U	a_1	a_2	a_3	a_4	a_5	d
1	85	100	98	100	90	1
2	89	97	64	20	73	1
3	90	67	63	17	49	3
4	88	86	63	15	58	1
5	72	80	76	40	81	2
6	78	79	76	36	78	2

$$(\sigma_i^k)=\begin{bmatrix}\sigma_1^1 & \sigma_1^2 & \sigma_1^3\\ \sigma_2^1 & \sigma_2^2 & \sigma_2^3\\ \sigma_3^1 & \sigma_3^2 & \sigma_3^3\\ \sigma_4^1 & \sigma_4^2 & \sigma_4^3\\ \sigma_5^1 & \sigma_5^2 & \sigma_5^3\\ \sigma_6^1 & \sigma_6^2 & \sigma_6^3\end{bmatrix}=\begin{bmatrix}0.945 & 0 & 0\\ 0.45 & 0.244 & 0.244\\ 0.220 & 0.293 & 0.307\\ 0.374 & 0.145 & 0.257\\ 0.275 & 0.433 & 0.2\\ 0.309 & 0.388 & 0.2\end{bmatrix}$$

Thus, we can obtain

$$\max_{1\le k\le 3}\{\sigma_1^k\}=\sigma_1^1=0.945,\quad \max_{1\le k\le 3}\{\sigma_2^k\}=\sigma_2^1=0.45,\quad \max_{1\le k\le 3}\{\sigma_3^k\}=\sigma_3^3=0.307,$$

$$\max_{1\le k\le 3}\{\sigma_4^k\}=\sigma_4^1=0.374,\quad \max_{1\le k\le 3}\{\sigma_5^k\}=\sigma_5^2=0.433,\quad \max_{1\le k\le 3}\{\sigma_6^k\}=\sigma_6^2=0.388.$$

6.3 The Grade of Grey Degree of Grey Numbers and Grey Membership Function Based on Rough Membership Function

Rough set theory can be regarded as a kind of extension of the classical set theory. From viewpoint of set-oriented theory, it defines a rough set with the application of a rough membership function, where the rough membership function can be explained by the conditional probability. One of the good properties of the model is the formulized representation and interpretation of the membership function and the operator of set theory embodied in the theory. According to the semantic definition of the grey degree of grey numbers and referencing normative definition and the idea of grade of the grey degree of grey numbers that was proposed by Professor Sifeng Liu, we give the definition of grey membership function and the grade of the grey degree of grey numbers based on rough membership function, and provide a standardized definition and semantic interpretation of the grey degree of grey numbers with a set-oriented view.

In rough set theory, rough approximation of set and rough membership function are closely connected with the grey degree of grey numbers. If $\mu_X(x) = 0$ or $\mu_X(x) = 1$, then an object x certainly belongs to a class X or not. In this case, the classification is certain, and the grey degree of grey numbers is minimum; if $\mu_X(x) > 0$ and $\mu_X(x) < 1$, then x belonging to X or not is in a grey transition state; if $\mu_X(x) = 0.5$, then the degree of certainty to which x belongs to X or not is maximally uncertain, and the grey degree of grey numbers is maximal; if the rough membership value of the input pattern approaches to either 0 or 1, the uncertainty about the belongingness of the pattern in the output class decreases, and the grey degree of grey numbers also decreases. Reversely, when $\mu_X(x)$ is close to 0.5, the degree of uncertainty about which x belonging to X or not increases, and the corresponding grey degree of grey numbers also increases.

Based on the above analysis, rough membership function is divided into upper rough membership function denoted by $\overline{\mu}_X(x)$ and lower rough membership function denoted by $\underline{\mu}_X(x)$, where $\overline{\mu}_X(x) \in [0.5, 1]$, $\underline{\mu}_X(x) \in [0, 0.5]$. Correspondingly, grey membership function is divided into upper rough membership function denoted by $\overline{g}_X(x)$ and lower rough membership function denoted by $\underline{g}_X(x)$. Obviously, upper rough membership function, lower rough membership function, and rough membership function take on the following properties:

1. $\overline{\mu}_X(x) = 1 - \underline{\mu}_X(x)$
2. $\mu_{X\cup Y}(x) = \mu_X(x) + \mu_Y(x) - \mu_{X\cap Y}(x)$
3. $\max(0, \mu_X(x) + \mu_Y(x) - 1) \leq \mu_{X\cap Y}(x) \leq \min(1, \mu_X(x) + \mu_Y(x))$

Hereby, grey membership function is defined based on rough membership function as follows.

Definition 6.16 Suppose x is an object of universe U, $x \in U$, X is a subset of universe U, $X \subseteq U$, $\overline{\mu}_X(x) \in [0.5, 1]$ and $\underline{\mu}_X(x) \in [0, 0.5]$. From U to closed interval [0, 1], two mappings are given by

$$\overline{\mu}_X(x): U \rightarrow [0.5, 1], \quad \mu \mid\rightarrow \overline{g}_X(x) \in [0, 1]$$

and

$$\underline{\mu}_X(x): U \rightarrow [0, 0.5], \quad \mu \mid\rightarrow \underline{g}_X(x) \in [0, 1]$$

then, $\overline{g}_X(x)$ and $\underline{g}_X(x)$ are, respectively, called the upper grey membership function and lower grey membership function.

Definition 6.17 Let $x \in U$, $X \subseteq U$, the level of grey degree of grey numbers to which an object x belongs to X or not is denoted by g_c; the upper grey membership function and the corresponding level of grey degree of grey numbers are, respectively, denoted by $\overline{g}_X(x)$ and $\overline{g}_c$, while the lower grey membership function and the corresponding level of grey degree of grey numbers are, respectively, denoted by $\underline{g}_X(x)$ and $\underline{g}_c$. Then, the levels of grey degree of grey numbers are defined as follows:

1. The grade of white numbers is defined as

$$g_c = 0: \underline{\mu}_X(x) = 0 \Rightarrow \underline{g}_c = 0 \quad \text{and} \quad \overline{\mu}_X(x) = 1 \Rightarrow \overline{g}_c = 0.$$

2. Grade 1 of grey numbers is defined as

$$g_c = 1 : \underline{\mu}_X(x) \in (0,0.1] \Rightarrow \underline{g_c} = 1 \quad \text{and} \quad \bar{\mu}_X(x) = [0.9,1) \Rightarrow \bar{g}_c = 1.$$

3. Grade 2 of grey numbers is defined as

$$g_c = 2 : \underline{\mu}_X(x) \in (0.1,0.2] \Rightarrow \underline{g_c} = 2 \quad \text{and} \quad \bar{\mu}_X(x) = [0.8,0.9) \Rightarrow \bar{g}_c = 2.$$

4. Grade 3 of grey numbers is defined as

$$g_c = 3 : \underline{\mu}_X(x) \in (0.2,0.3] \Rightarrow \underline{g_c} = 3 \quad \text{and} \quad \bar{\mu}_X(x) = [0.7,0.8) \Rightarrow \bar{g}_c = 3.$$

5. Grade 4 of grey numbers is defined as

$$g_c = 4 : \underline{\mu}_X(x) \in (0.3,0.4] \Rightarrow \underline{g_c} = 4 \quad \text{and} \quad \bar{\mu}_X(x) = [0.6,0.7) \Rightarrow \bar{g}_c = 4.$$

6. Grade 5 of grey numbers is defined as

$$g_c = 5 : \underline{\mu}_X(x) \in (0.4,0.5] \Rightarrow \underline{g_c} = 5 \quad \text{and} \quad \bar{\mu}_X(x) = [0.5,0.6) \Rightarrow \bar{g}_c = 5.$$

7. The grade of black numbers is defined as

$$g_c > 5 : \underline{\mu}_X(x) = \bar{\mu}_X(x) \quad \underline{g_c} = \bar{g}_c > 5.$$

The concept chart with the levels of grey degree of grey numbers based on rough membership function is depicted in Figure 6.9.

If $\mu_X(x) \in [0, 1]$, then $\underline{g}_X(x) = 0$ and $\bar{g}_X(x) = 1$. In this case, no uncertain information is associated with the object x. Hence, the grey degree of grey numbers is named grey degree of white numbers, that is, $g_c = \bar{g}_c = \underline{g}_c = 0$; if $\mu_X(x) = 0.5$, then $\underline{g}_X(x) = \bar{g}_X(x) = 1$, the degree of uncertainty whether x belonging to X or not is maximal. Hence, the grey degree of grey numbers is named grey degree of black numbers. In terms of Definition 6.2.2, the higher the level of grey degree of grey numbers is, the vaguer the information granularity is. On the contrary, the lower the level of grey degree of the grey numbers is, the clearer the information granularity is.

As $\bar{\mu}_X = 1 - \underline{\mu}_X(x)$, if the degree of certainty for x belonging to X is denoted by the level of upper grey degree of grey numbers, and the degree of certainty for x not belonging to X is denoted by the level of lower grey degree of grey numbers, then the two relations are complementary.

According to Definition 6.17, because the level of grey degree of grey numbers is decided by the interval where possible maximum rough membership degree is in, the whitened values of the grey numbers for $\bar{g}_c$ and $\underline{g}_c$ are, respectively, defined

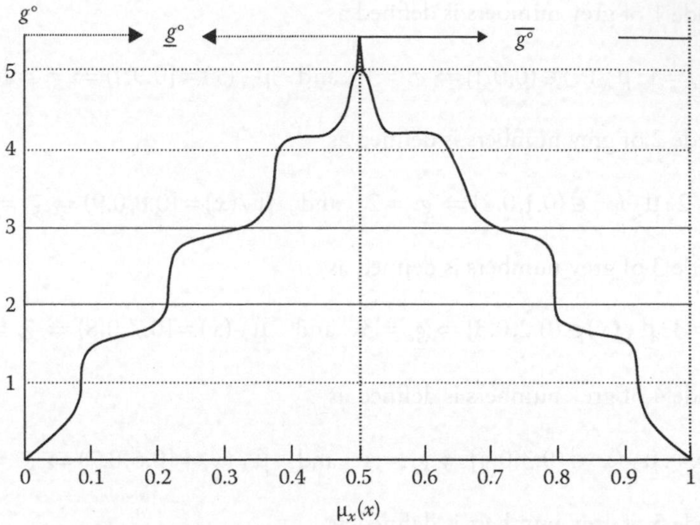

Figure 6.9 Concept chart for the level of grey degree of grey numbers.

as the possible maximum rough membership degree of the corresponding interval. For example, if we can calculate the possible maximum rough membership degree $\mu_X(x) = 0.75$ of a subset of conditions subset according to the decision table, due to $0.75 \in [0.7, 0.8]$, then $\mu_X(x) = 0.75$ is the whitened value for the grade of the upper grey degree of grey numbers $\overline{g}_c = 3$.

6.4 Grey Rough Approximations

Definition 6.18 Suppose that $S = (U, A, V, f)$, $A = C \cup D$, $X \subseteq U$, $P \subseteq C$, at a specified level $g_c \leq 5$, $\underline{g}_c$-lower approximation, and $\overline{g}_c$-upper approximation of X with I_p in S, are defined, respectively, by

$$\underline{apr}_P^{g_c}(X) = \bigcup \left\{ \frac{|I_P(x) \cap X|}{|I_P(x)|} \leq \overline{g}_c \right\} \tag{6.19}$$

$$\overline{apr}_P^{g_c}(X) = \bigcup \left\{ \frac{|I_P(x) \cap X|}{|I_P(x)|} > \underline{g}_c \right\} \tag{6.20}$$

where the level of upper grey degree of grey numbers $\overline{g}_c$ is corresponding with the upper grey membership function $\overline{\mu}_X(x) \in (0.5, 1]$, while the level of the lower grey degree of grey numbers $\underline{g}_c$ is corresponding with the lower grey membership function $\mu_X(x) \in [0, 0.5)$.

Under the level of grey degree of grey numbers g_c, the $\underline{g}_c$-lower approximation of the set $X \subseteq U\, \underline{apr}_P^{g_c}(X)$ is made up of the unions of all equivalence classes belonging to X whose level of grey degree of grey numbers are below or equal to the level of upper grey degree of grey numbers $\overline{g}_c$ in U, while the $\underline{g}_c$-lower approximation consists of the unions of all equivalence classes belonging to X whose level of grey degree of grey numbers are above the level of lower grey degree of grey numbers $\underline{g}_c$.

Definition 6.19 Suppose $S = (U, A, V, f)$, $A = C \cup D$, $X \subseteq U$, $P \subseteq C$, at a specified level $g^\circ \leq 5$, the measure of classification quality is defined by

$$\gamma_P^{g_c}(P, D) = \frac{\left| \cdot \left\{ \frac{|X \cap I_P(x)|}{|I_P(x)|} \leq \overline{g}_c \right\} \right|}{|U|} \tag{6.21}$$

The value $\gamma_P^{g_c}(P, D)$ measures the proportion of objects in the universe in which classification is possible at the specified level, $g^\circ \leq V$. While approximate reduct $red_P^{g_c}(C, D)$ is the minimal set of condition attributes that does not contain redundant attributes and ensures a clear classification at the specified level $g^\circ \leq V$.

The classification of objects in boundary region is uncertain in rough set, and whether the objects in boundary region can be classified into a specified class is mostly decided by the given level of grey degree of grey numbers. Thus, the grey rough approximations defined are similar to the definition of variable precision rough set. If the given level $\overline{g}_c$ and $\underline{g}_c$ are, respectively, the corresponding whitened values, then grey rough approximation is changed to rough approximation in the sense of variable precision rough set. Consequently, variable precision rough approximation can be viewed as a special case of grey rough approximation. Compared with variable precision rough set theory, whether an object in relatively rough set X is discernable depends on a confident threshold value β. If any β-value chosen is equal or below the upper threshold, the set X is discernable. Otherwise, there is no opportunity for majority to be included. Hence, the set X is indiscernible. In real situation, it is generally difficult to decide the possible maximum confident threshold value on the β-value in advance, especially for large database. The concept of the interval grey number provides a real numeric tool by giving lower endpoint and upper endpoint, and the representation is very useful in the case that we cannot obtain the exact measure of real amount. Further, for some applications, what the decision-makers concern is the degree of certainty that the event of interest would occur, or would not occur, rather than to predict the occurrence of the event by a high probabilistic rule. In such cases, how to set the confident threshold value is not so important. For example, in medical domain, the results of medical test might indicate increased/decreased opportunities of a specific disease. Yet, it is impossible to predict the probability the disease would happen at a specified confident threshold value. In such applications, it is more reasonable to generate rule set by setting the level of grey degree of grey numbers instead of confident threshold value.

Proposition 6.5 If $g_c = 0$, then grey rough approximation degrade to the rough approximation of initial rough set.

Proof: If $g_c = 0$, $\mu_X(x) \in [0, 1]$, then

$$\underline{apr}_P^{g_c}(X) = \bigcup\left\{\frac{|I_P(x) \cap X|}{|I_P(x)|} = 1\right\}$$

thus, $I_P(x) \subseteq X$ can be obtained. The lower approximation of grey rough set is changed to the lower approximation of initial rough set. As for

$$\overline{apr}_P^{g_c}(X) = \bigcup\left\{\frac{|I_P(x) \cap X|}{|I_P(x)|} > 0\right\}$$

we can obtain $I_P(x) \cap X \neq \phi$, the upper approximation of grey rough set is transformed to the upper approximation of initial rough set. In other words, original rough set model can be regarded as a special case of grey rough set model in the case of clear information granularity.

For the inconsistent rules in the sense of initial rough set, the inconsistency degree is weak according to the setting level of grey degree of grey numbers g_c, thus an inconsistent rule can be considered because of grey properties from some noises mixed in the given data. This rule or the main part of it can be viewed as deterministic one. However, if the level of grey degree of grey numbers is high, that is to say, the inconsistency degree is strong, and then the corresponding rule is real indeterministic and should be treated as a random rule.

Proposition 6.6 For $g_c \leq 5$, $X \subseteq U$ and $Y \subseteq U$, the following holds:

1. $\overline{apr}_P^{g_c}(X \cup Y) \supseteq \overline{apr}_P^{g_c}(X) \bigcup \overline{apr}_P^{g_c}(Y)$
2. $\underline{apr}_P^{g_c}(X \cap Y) \subseteq \underline{apr}_P^{g_c}(X) \bigcap \underline{apr}_P^{g_c}(Y)$
3. $\underline{apr}_P^{g_c}(X \cup Y) \supseteq \underline{apr}_P^{g_c}(X) \bigcup \underline{apr}_P^{g_c}(Y)$
4. $\overline{apr}_P^{g_c}(X \cap Y) \subseteq \overline{apr}_P^{g_c}(X) \bigcap \overline{apr}_P^{g_c}(Y)$

Proof

1. For any $X \subseteq U$ and $Y \subseteq U$, given $g_c \le 5$, then,

$$\frac{|I_P(x) \cap (X \cup Y)|}{|I_P(x)|} \ge \frac{|I_P(x) \cap X|}{|I_P(x)|} \text{ and } \frac{|I_P(x) \cap (X \cup Y)|}{|I_P(x)|} \ge \frac{|I_P(x) \cap Y|}{|I_P(x)|}$$

therefore, $\overline{apr}_P^{g_c}(X \cup Y) \supseteq \overline{apr}_P^{g_c}(X) \bigcup \overline{apr}_P^{g_c}(Y)$.

2. For any $X \subseteq U$ and $Y \subseteq U$, given $g_c \le 5$, then,

$$\frac{|I_P(x) \cap (X \cap Y)|}{|I_P(x)|} \le \frac{|I_P(x) \cap X|}{|I_P(x)|} \text{ and } \frac{|I_P(x) \cap (X \cap Y)|}{|I_P(x)|} \le \frac{|I_P(x) \cap Y|}{|I_P(x)|}$$

therefore, $\underline{apr}_P^{g_c}(X \cap Y) \subseteq \underline{apr}_P^{g_c}(X) \bigcap \underline{apr}_P^{g_c}(Y)$.

Equations 6.3 and 6.4 can be proved in the same way.

Proposition 6.7 $\underline{apr}_P^{g_c}(X) \subseteq \overline{apr}_P^{g_c}(X)$.

Proof: Let $x \in \underline{apr}_P^{g_c}(X)$, because the equivalence relation is reflexive, $\forall x \in I_P(x)$. According to Definition 6.16, for $g_c \le 5$, the rough membership degree in the interval of $\overline{g}_c$ is greater than the one in $\underline{g}_c$, hence, $x \in \overline{apr}_P^{g_c}(X)$. That is $\underline{apr}_P^{g_c}(X) \subseteq \overline{apr}_P^{g_c}(X)$.

Example 6.7 Compute the reduct at different levels of grey degree of grey numbers and the corresponding rule set according to the decision tables in Example 3.4.

Solution: The universe U is partitioned into equivalence classes as follows:

$$\frac{U}{C} = \{X_1, X_2, X_3, X_4, X_5\}$$

where $X_1 = \{n_1\}$, $X_2 = \{n_2\}$, $X_3 = \{n_3, n_4\}$, $X_4 = \{n_5\}$, $X_5 = \{n_6\}$;

$$\frac{U}{D} = \{Y_A, Y_D, Y_C\}$$

where $Y_A = \{n_1, n_2, n_5\}$, $Y_D = \{n_3, n_4\}$, $Y_C = \{n_6\}$.

According to the proposed method in the paper, the reduct at different levels of grey degree of grey numbers and the corresponding rule sets obtained are in Table 6.4.

Table 6.4 g_c-Reduct and the Corresponding Rule Sets Generated

Level	*g_c-Reduct*	*Rules*	*Support*	*Classification Quality*
$g_c = 1$	$\{a_1, a_2, a_3\}$	$a_1 = \text{No} \wedge a_2 = \text{Normal} \wedge a_3 = \text{Megalo} \rightarrow d = \text{Aarskog}$	1	$\gamma_P^{g^c}(P,D) = 1$
		$a_1 = \text{Yes} \wedge a_2 = \text{Hyper} \wedge a_3 = \text{Megalo} \rightarrow d = \text{Aarskog}$	1	
		$a_1 = \text{Yes} \wedge a_2 = \text{Hyper} \wedge a_3 = \text{Normal} \rightarrow d = \text{Down}$	2	
		$a_1 = \text{Yes} \wedge a_2 = \text{Hyper} \wedge a_3 = \text{Large} \rightarrow d = \text{Aarskog}$	1	
		$a_1 = \text{No} \wedge a_2 = \text{Hyper} \wedge a_3 = \text{Megalo} \rightarrow d = \text{Cat} - \text{cry}$	1	
$g_c = 3$	$\{a_4, a_6\}$	$a_4 = \text{Yes} \wedge a_6 = \text{Long} \xrightarrow{g^\circ = 3} d = \text{Aarskog}$	4	$\gamma_P^{g^c}(P,D) = 1$
		$a_4 = \text{No} \wedge a_6 = \text{Normal} \rightarrow d = \text{Down}$	2	
$g_c = 4$	$\{a_2, a_4\}$	$a_2 = \text{Normal} \wedge a_4 = \text{Yes} \rightarrow d = \text{Aarskog}$	1	$\gamma_P^{g^c}(P,D) = 1$
		$a_2 = \text{Hyper} \wedge a_4 = \text{Yes} \xrightarrow{g^\circ = 4} d = \text{Aarskog}$	3	
		$a_2 = \text{Hyper} \wedge a_4 = \text{No} \rightarrow d = \text{Down}$	2	

6.5 Reduced Attributes Dominance Analysis Based on Grey Correlation Analysis

In rough set theory, the reduced attributes are usually regarded as being of equal importance, and grey correlation analysis can be utilized to further make dominance analysis on the reduced conditional attributes sets.

Example 6.8 Take the reduct $\{a_2, a_4, a_6, a_8\}$ of the post-evaluation of construction projects in Chapter 3, for example, which are shown in Table 6.5, where a_2, evaluation of construction period management; a_4, evaluation of technology; a_6, evaluation of project construction quality; a_8, evaluation of environmental protection; and d, evaluation of comprehension. Make dominance analysis on the reduct.

Table 6.5 A Decision Table for Process Evaluation of Construction Project

U	a_2	a_4	a_6	a_8	d
n_1	92.20	88.23	95.67	90.45	91.49
n_2	84.21	92.56	87.30	88.23	88.32
n_3	94.37	98.51	95.79	96.63	96.64
n_4	80.20	70.43	64.60	70.36	69.73
n_5	67.27	68.26	69.63	72.60	69.55
n_6	82.45	83.15	73.33	76.30	78.39
n_7	93.12	86.12	91.13	86.17	89.81
n_8	83.70	76.31	80.70	74.26	79.30
n_9	98.16	97.68	96.61	86.63	93.22
n_{10}	93.75	93.34	97.90	91.70	91.80
n_{11}	83.46	86.33	80.20	85.71	82.97
n_{12}	92.14	88.40	86.53	80.79	87.69
n_{13}	95.20	96.23	93.21	86.30	91.42
n_{14}	86.23	90.67	90.62	83.46	87.38
n_{15}	81.90	81.21	81.79	72.43	79.51
n_{16}	82.63	81.56	83.40	78.91	79.26
n_{17}	89.65	90.43	91.73	93.67	90.48
n_{18}	79.21	86.81	88.55	87.40	83.28

Solution

1. Calculate the initial value image (or mean value image) of different sequences. Let $A_i' = A_i/a_i(1) = (a_i'(1), a_i'(2), \ldots, a_i'(5))$, $i = 0, 1, 2, \ldots, 18$, the calculated results are listed in Table 6.6.
2. Compute minus sequences. $\Delta_i(k) = |a_i'(k) - a_i'(5)|$, where $i = 1, 2, \ldots, 18$, $k = 1, 2, \ldots, 5$, the calculated results are listed in Table 6.7.
3. Calculate the maximum minus value and the minimum minus value. Let $M = \max_i \max_k \Delta_i(k) = 0.1077$, $m = \min_i \min_k \Delta_i(k) = 0$.
4. Compute correlation coefficient Let $M = \max_i \max_k \Delta_i(k) = 0.1077$, $m = \min_i \min_k \Delta_i(k) = 0$, $\xi = 0.5$, then

Table 6.6 An Initial Value Image of Different Sequences

a_2	a_4	a_6	a_8	d
1.0000	1.0000	1.0000	1.0000	1.0000
0.9133	1.0491	0.9125	0.9755	0.9654
1.0235	1.1165	1.0013	1.0683	1.0563
0.8698	0.7983	0.6752	0.7779	0.7622
0.7296	0.7737	0.7278	0.8027	0.7602
0.8943	0.9424	0.7665	0.8436	0.8568
1.0100	0.9761	0.9525	0.9527	0.9816
0.9078	0.8649	0.8435	0.8210	0.8668
1.0646	1.1071	1.0098	0.9578	1.0189
1.0168	1.0579	1.0233	1.0138	1.0034
0.9052	0.9785	0.8383	0.9476	0.9069
0.9993	1.0019	0.9045	0.8932	0.9585
1.0325	1.0907	0.9743	0.9541	0.9992
0.9352	1.0277	0.9472	0.9227	0.9551
0.8883	0.9204	0.8549	0.8008	0.8691
0.8962	0.9244	0.8717	0.8724	0.8663
0.9723	1.0249	0.9588	1.0356	0.9890
0.8591	0.9839	0.9256	0.9663	0.9103

Table 6.7 Minus Sequences

0.0000	*0.0000*	*0.0000*	*0.0000*	*0.0000*
0.0520	0.0837	0.0528	0.0101	0.0000
0.0328	0.0602	0.0550	0.0120	0.0000
0.1077	0.0361	0.0869	0.0157	0.0000
0.0306	0.0135	0.0324	0.0425	0.0000
0.0374	0.0856	0.0903	0.0133	0.0000
0.0283	0.0056	0.0291	0.0290	0.0000
0.0410	0.0019	0.0232	0.0458	0.0000
0.0457	0.0882	0.0091	0.0611	0.0000
0.0134	0.0545	0.0199	0.0104	0.0000
0.0017	0.0716	0.0686	0.0407	0.0000
0.0409	0.0435	0.0540	0.0653	0.0000
0.0333	0.0914	0.0249	0.0451	0.0000
0.0198	0.0726	0.0079	0.0324	0.0000
0.0192	0.0514	0.0141	0.0683	0.0000
0.0299	0.0581	0.0054	0.0061	0.0000
0.0166	0.0360	0.0301	0.0466	0.0000
0.0512	0.0736	0.0153	0.0560	0.0000

$$\gamma_{0i}(k) = \frac{m + \xi M}{\Delta_i(k) + \xi M} = \frac{0.0538}{\Delta_i(k) + 0.0538}$$

hence, the grey incidence coefficients are listed in Table 6.8.

5. Calculate degree of grey correlation

   $\gamma_{5i}(k) = (1/n)\sum_{k=1}^{n} \gamma_{0i}(k) = (1/18)\sum_{i=1}^{18} \gamma_{0i}(k)$, grey degree of grey incidence calculated is $\gamma_{51}(k) = 0.6553$, $\gamma_{52}(k) = 0.5660$, $\gamma_{53}(k) = 0.6625$, $\gamma_{54}(k) = 0.6580$.
6. Analysis of results

   On account of $\gamma_{53}(k) > \gamma_{54}(k) > \gamma_{51}(k) > \gamma_{52}(k)$, for the decision attribute "comprehensive evaluation," that is, project construction quality > environmental protection > construction period management > technology in the reduct $\{a_2, a_4, a_6, a_8\}$.

Table 6.8 Grey Incidence Coefficients

a_2	a_4	a_6	a_8	d
1.0000	1.0000	1.0000	1.0000	1.0000
0.5085	0.3912	0.5045	0.8419	1.0000
0.6216	0.4718	0.4943	0.8172	1.0000
0.3332	0.5985	0.3823	0.7738	1.0000
0.6376	0.7998	0.6243	0.5589	1.0000
0.5897	0.3859	0.3733	0.8023	1.0000
0.6550	0.9065	0.6490	0.6501	1.0000
0.5672	0.9665	0.6984	0.5404	1.0000
0.5405	0.3789	0.8555	0.4681	1.0000
0.8003	0.4966	0.7298	0.8376	1.0000
0.9699	0.4291	0.4396	0.5692	1.0000
0.5682	0.5531	0.4991	0.4519	1.0000
0.6177	0.3704	0.6832	0.5439	1.0000
0.7307	0.4257	0.8725	0.6244	1.0000
0.7367	0.5115	0.7919	0.4407	1.0000
0.6429	0.4809	0.9084	0.8983	1.0000
0.7640	0.5993	0.6409	0.5356	1.0000
0.5126	0.4222	0.7784	0.4899	1.0000

6.6 Summary

As the simulation and processing tools for incomplete information, grey system and rough set have been developed into far-reaching soft technological methodology being dedicated to analyzing uncertain and incomplete data. Taking the uncertain systems of the partially known and partially unknown information with small sample and poor information as the research object, and extracting valuable information by mainly producing or mining the partially known information, the grey system theory can properly describe and effectively control the behavior and the law of system. Grey system theory has been successfully applied to the prediction of many important projects, and the predictive precision is high. Whereas the theoretic basis of grey system has some shortages, such as the definition of grey degree of grey numbers and the explanation for the relations between grey degrees of different grey numbers are some disputed topics. According to the advantages and disadvantages of the grey system theory and rough set theory, this chapter studies the hybrid of the two methods.

Rough membership of rough sets theory is obtained by directly calculating the data analyzed and can objectively describe the degree of uncertainty of the data. In the chapter, a formal definition and the grade of grey degree of grey numbers are proposed based on the rough membership function, which explicitly states the levels of grey degree of grey number for the classification of objects and explicitly interprets the semantics of grey degree of grey numbers. Besides, the whitened values in different levels of grey numbers are given. These can complement the shortages of definition of grey degree of grey numbers in grey systems, while grey rough approximation constructed can classify the objects of boundary region by setting the levels of grey degree of grey numbers, which makes up for the insufficiency of the definition of rough set. In addition, grey clustering can provide a method for rough set to construct decision tables, and grey correlation analysis can be utilized to further compare relative dominance of the reduced conditional attributes.

Chapter 7

A Hybrid Approach of Variable Precision Rough Sets, Fuzzy Sets, and Neural Networks

Neural network is a technique of artificial intelligence widely used in pattern recognition and machine learning. It possesses strong adaptability and learning ability, and its precision in the data classification and prediction is extremely high. Being different from classical statistical methods, the neural network can be applied to data analysis of small sample, without satisfying normal distribution. Because of the nonlinear transformation involved in the radbas layer, neural networks can work well in imitating complex data. However, neural networks are very complex in the computation, and the time required by their learning algorithms is in most cases greater than that required by other methods in artificial intelligence. Besides, it cannot provide a crisp description of its discovery pattern. Variable precision rough set (VPRS) has the advantage that we can get learning probability decision rules from imprecision data and that it can be expressed by the customer's understanding. The hybrid approach of VPRSs, fuzzy sets, and neural network can shorten the time of network learning and improve the prediction ability of data and strengthen explanatory power.

7.1 Neural Network

Artificial neural network (ANN) is an active cross-discipline involving biology, mathematics, physics, electronics, and computer technology, and has an extensive application prospect. As is well known, human beings' brains possess extremely perfect and strong functions such as memory, calculation, analysis, judgment, and decision. Although the understanding about the detailed mechanism of our brain's thinking activities is far from complete, people have already known its basic structure and functions for a long time. The brain is made up of a large number of neural cells. Although a single cell has only one basic function, it is extensively connected with other cells in a whole and can complete quickly various complex and difficult tasks under the mechanism effect of "biology, electronics, and chemistry." ANN adopts physical visible system to imitate the structure and function of the brain, and it is made up of extensively connected processing units. Each structure and function of the processing unit is simple, and it usually completes some basic transformation and the working mode of the whole system is greatly different from the present computers with program instruction by serial mode.

The aim of studying neural network is to simulate knowledge acquisition and organizational skills of human being's brain, and it can establish its position as an important nonlinear classification and high adaptive system. Neural network can perform the specific reasoning tasks, such as pattern classification, pattern cluster, pattern image, pattern association, and feature mapping.

Compared with the classical methods, the advantages of the neural network method are as follows:

1. There exists strong error-tolerance, and it is able to recognize input pattern with noise or deformation.
2. There exists a strong self-adoption learning ability.
3. It can recognize quickly with the abilities of parallel-distributed information storage and processing.
4. It can combine recognition processing with pretreatment.

Neural network can achieve high prediction precision without any hypothesis in pattern recognition of input data. Therefore, for neural network technology, some application achievements have been reached in financial fields that require pattern match, pattern classification, and pattern prediction in such occasions as bankruptcy prediction, borrowing and lending ability evaluation, and prediction of credit rank. However, neural network is regarded as "black box," and it is difficult to interpret the results of operation. With the increase of dimensionality of the problems being tackled, the computational complexity becomes visible as the sizes of the datasets to be handled grow very quickly, and many scholars have studied the learning process of accelerating neural network.

7.1.1 An Overview of the Development of Neural Network

ANN was first proposed by the American psychologist, McCulloch, and mathematical logician, Pitts. The input and output relationship of its neurons is defined as follows:

$$y_j = sign\left(\sum_i w_{ji}\, x_i - \theta_j\right) \tag{7.1}$$

where the input and output are both binary value, and w_{ji} is a fixed weight value. Some logical relationship can be realized by taking advantage of a simple network. Although this model is too simple, it lays the foundation for further research.

In 1949, the American psychologist, Hebb, proposed the rules of connection change between neurons according to the mechanism of conditional reflex in psychology. The basic idea of the rules is that when two neurons excite or inhibit simultaneously, the connecting strength between them will increase, and it can be illustrated as follows:

$$w_{ji} = \begin{cases} \sum_{k=1}^{n} x_i^{(k)} x_j^{(k)} & i \neq j \\ 0 & i = j \end{cases} \tag{7.2}$$

The significance of the learning rule is that the adjust of the connecting weight is proportional to the product of two neurons in active state, and the connecting weight is symmetric and the connecting weight from neuron to itself is "0." There are still many learning rules adopted by neural networks.

In the 1950s, the American scholar, Rosenblatt, proposed the perceptron model, and he thought that perceptron model was the simplified model for biologic system to percept the outside information and was mainly used in pattern classification. In the 1960s, Miskey and Papert pointed out in their works that the function of simple linear perceptron was limited and it was unable to solve the linear and inseparable classification problems that contain two kinds of samples. The typical example is the operation of "not" or "or," and, that is to say, the simple linear perceptron cannot realize the logical relationship of "not" or "or." Radbas node must be added to solve this problem. However, for multilayer network, finding an effective learning algorithm is still a problem that is difficult to solve. Therefore, the study of the neural network was in low tide in the 1970s.

In the 1980s, the second vigorous campaign of neural network rose. Hopfield proposed a kind of feedback interconnection network, which was called Hopfield network. He defined an energy function that is a function of neuron's state and connecting weight value, and we can solve problems of associational memory and perfect calculation by making use of the network, while the most typical case is that it solves successfully the problem about the most perfect choice for a traveling businessman. Rumelhart and McClelland put forward the back propagation of multilayer feedforward network, which was called BP algorithm for short. It solved the problems that perceptron could not solve in the past. The creative research work by these scholars promoted the rapid development of ANN's research.

Neural network is put forward based on the achievement of the modern neural science research. It reflects some characteristics of brain's function, although it is not the real description of neural system but its simplification, abstract, and simulation. That is to say, ANN is an abstract math model, which can be used as brain's structure model recognition model and computer information processing methods or algorithm structure from different research's points of view. Therefore, the directions of neural network research can be classified into three kinds:

1. Probe into the biologic structure and mechanism of brains neural network, which is the initial purpose of neural network theory.
2. Use neural network theory as means and methods to solve some problems that cannot be solved or are difficult to solve with the classical methods.
3. Use special function network formed by microelectronics or optics parts, which is the main research direction of the new generational computer-producing field.

7.1.2 Structure and Types of Neural Network

Neural network adopts the mode with realizable parts or computers through hardware or software to simulate some structure and function of organism nervous system to solve machine intelligent problems such as machine learning, recognition, control, and decision. A single calculation unit making up a neural network is called node, unit, or neuron. Generally speaking, neural network is a parallel and distributed information processing network structure, and many parallel operation functions of neural network are made up of simple neurons, which are similar to the neurons of a biological neural network system. Neural network is a nonlinear dynamics system and its unique feature lies in the distributed storage and parallel processing of information. There are many kinds of neuron models; a simple neuron model is shown in Figure 7.1.

Neural network is generally made up of many neurons, and every neuron has only one output, although it can be connected with other neurons. Every neuron input has many connecting passage and every passage corresponds to a connecting weight coefficient. The collection of simultaneous or parallel operation is called layer, and it can be classified into input layer, output layer, and

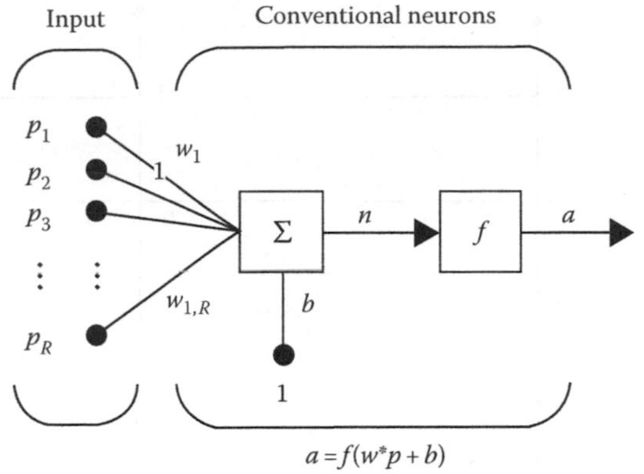

Figure 7.1 A neuron model.

radbas layer. Strictly speaking, neural network is a direction picture possessing some properties as follows:

1. Every node has state vector x_j.
2. There is a connecting weight coefficient w_{ji} between node i and j.
3. Every node has a threshold θ_j.
4. Every node can be defined a transfer function $f_j[x_i, w_{ji}, \theta_j(i \neq j)]$, and its common form is $f\left(\sum_i w_{ji}x_i - \theta_j\right)$. Some main transfer functions of neurons are illustrated in Figure 7.2.

Figure 7.2a and b state that the output value is a binary vector, if x_i is also a binary vector, the neural network is called discrete neural network; Figure 7.2c states that the input and output value is piecewise linear relationship; Figure 7.2d is a monotone increasing nonlinear smooth function, and the sigmoid can be presented as follows:

$$y_i = th\left(\frac{n_i}{n_0}\right) \tag{7.3}$$

$$y_i = \frac{1}{1+e^{-\beta n_i}} \approx \frac{1}{2}\left(1 + th\left(\frac{n_i}{n_0}\right)\right) \quad (\beta > 0) \tag{7.4}$$

In formula (7.3), $-1 \leq y_i \leq 1$, the curve becomes steep when it passes "0" with the reduction of parameter n_0; when $n_0 \to 0$, $th(n_i/n_0) \to \text{sgn}(n_i)$. While in formula (7.4), $-1 \leq y_i \leq 1$, when $\beta \to \infty$, y_i comes close to step function, and β is selected as 1 in common situation. So the curve presented by formula (7.4) is actually the up-moving and half-reduced amplitude of that presented by formula (7.3).

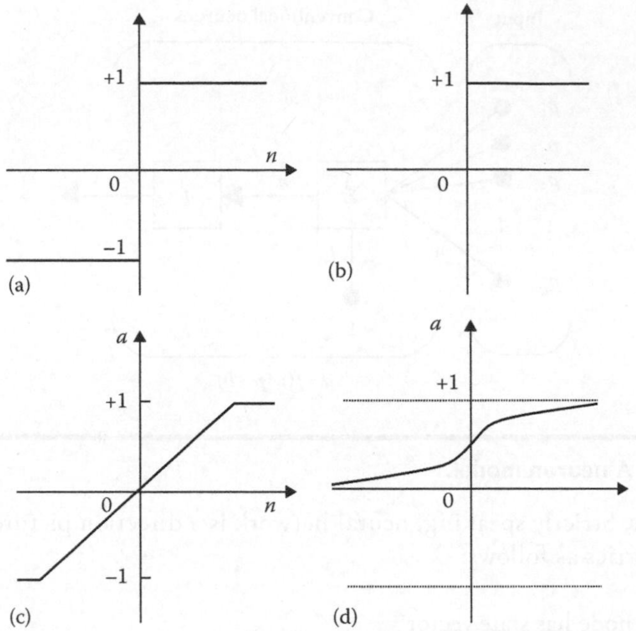

Figure 7.2 Four kinds of transfer functions of neuron: (a) sign function, (b) step function, (c) piecewise linear function, and (d) nonlinear monotone increasing function.

Although the structure and function of a single neuron is simple and limited, the behaviors that network system consisted of large number of neurons can realize are multiple. Compared with digital computer, neural network has collective computation ability and self-adaptive ability. In addition, it is also strong, fault toleranant, and robust, and is good at association, synthesis, and generalization.

There are various neural network models that can describe and simulate the different levels of biological neural system from different aspects. Some representative network models are perceptron, multilayer mapping BP network, radial basis function (RBF) network, RAM, Hopfield model, and probability neural network. We can realize the functions such as function approach, data cluster, pattern classification, and optimization with these network models.

7.1.3 Perceptron

Perceptron is one of the earliest proposed kind of neutral network model. It is a neutral network with one-layer neuron, and the activation function is linear total threshold unit. The initial perceptron algorithm only has one output joint, and it equals to one neuron. When it is applied to two kinds of patterns classification, it resembles the situation that two sorts of samples are separated by one hyperplane in the space of high dimensional sample. Rosenblatt has proved that the algorithm must be convergent if two sorts of patterns are separable (refers to one hyperplane that can separate them apart). Perceptron is especially suitable for simple pattern classification problems.

7.1.3.1 Perceptron Neuron Model

A perceptron neural model is shown in Figure 7.3. The model has R input vectors, and `hardlim(·)` is used as the transfer function.

Every input $p_i(i = 1, 2,..., R)$ of perceptron neuron corresponds a weight value $w_{1,j}(j = 1, 2,..., R)$, and takes the sum of the product of all inputs and their correspondent weights just as one input of sigmoid transfer function, while another input is the product of network input constant 1 and threshold b. If $n = Wp + b$ is taken as input, then network output will acquire two cases that are 0 and 1, so that input space can be divided by perceptron neuron into two parts. Sigmoid transfer function is followed in Figure 7.4.

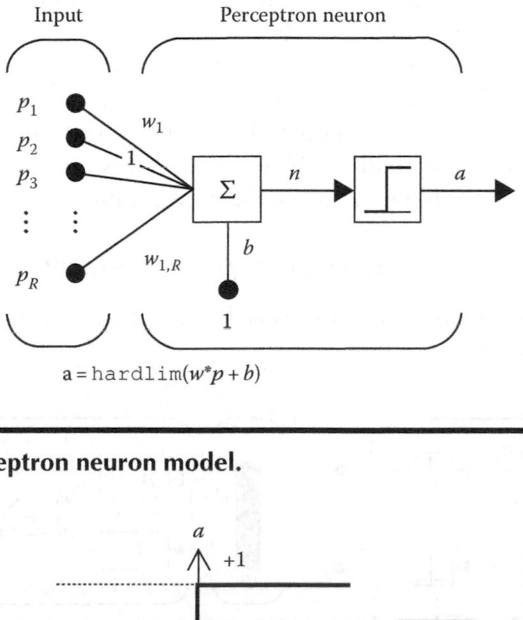

Figure 7.3 Perceptron neuron model.

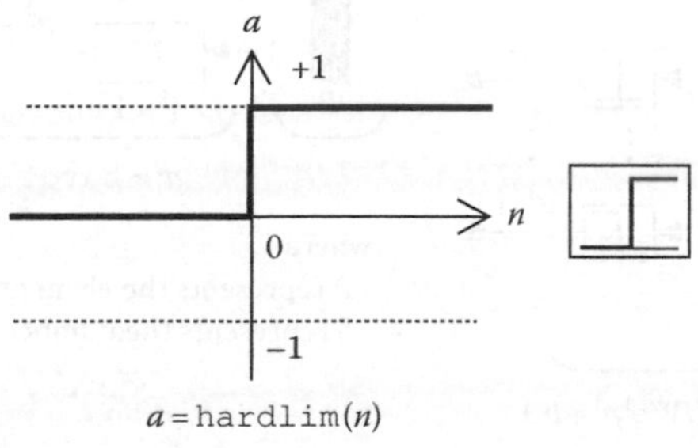

Figure 7.4 Sigmoid transfer function.

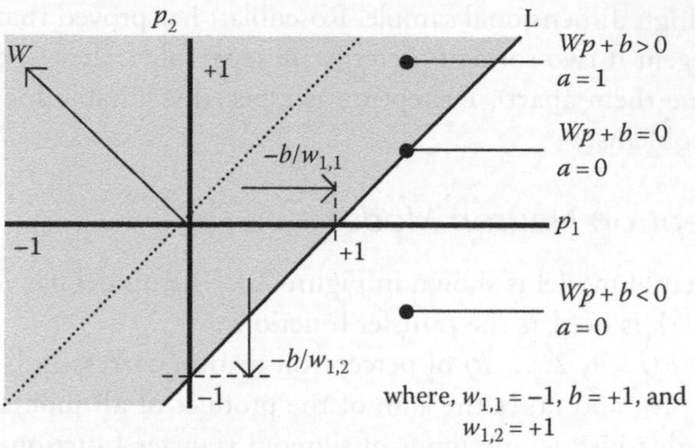

Figure 7.5 Classification results of input vectors.

If we take a signal-layer perceptron with two inputs whose weights are $w_{1,1} = -1$ and $w_{1,2} = 1$, and the threshold $b = 1$, for example, we can obtain the following classification results shown in Figure 7.5.

Dividing line L divides input space into two parts, and the input vectors above the straight line make the output of perceptron be 1, while the input vectors below the straight line make the output of perceptron be 0. In addition, the position of L can be controlled both by W and b, and this can make the perceptron distinguish two kinds of input vectors of the different positions.

7.1.3.2 Network Structure of Perceptron Neutral Network

Single-layer perceptron neutral network can be described in two different ways and is shown in Figure 7.6.

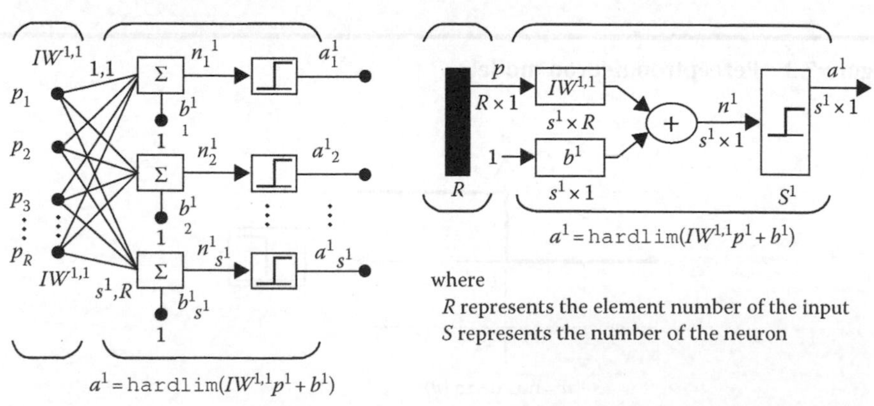

Figure 7.6 Network structure of single-layer perceptron.

In Figure 7.6, networks with R inputs connect S neurons by the weight value w_{ij}, while i and j show that weight value w_{ij} connects ith neuron and jth input, finally we can obtain S outputs.

From Figure 7.6, we can find that the neutral network of perceptron has only one-layer neuron, and it is caused by learning rules of perceptron that can only train one-layer network, while the structure limitation of perceptron neutral network restricts its application area to some extent.

7.1.3.3 Learning Rules of Perceptron Neutral Network

Learning rules is the algorithm used to count weight value of the neutral network and threshold, and we can adjust the weight and threshold of network by adjusting the algorithm so that it can make network output achieve target expectation values. The learning rules of perceptron belong to supervising learning rules. Supervising learning rules need the sample sets of input and output of network. Once a sample is put in, comparing output of network correspondingly with that of sample, we can adjust the weight and threshold of network through their difference, and make output of network furthest close to sample output of network, which is what we expect.

As to the perceptron whose input is p, output vector is a, target vector is t, the learning error is e, while $e = t - a$. At the moment, the revised formulas of weight value and threshold value are as follows:

$$\Delta IW_{i,j} = [t_i - a_i] * p_j = e_i * p_j$$

$$\Delta b_i = [t_i - a_i] * l = e_i * l$$

where $i = 1, 2, 3, \ldots, s$, $j = 1, 2, 3, \ldots, R$, the renewal weight and threshold are

$$b_i = b_i + \Delta b$$

$$IW_{i,j} = IW_{i,j} + \Delta IW_{i,j}$$

It can be expressed by the form of matrix as follows:

$$IW = IW + e * p^T$$

$$b = b + e$$

The algorithm has already been proved. If optional weight and threshold exists, then this algorithm can converge to optional weight and threshold after finite order cyclic.

7.1.4 Back Propagation Network

As the transfer function of neuron in perceptron neutral network adopts sign function, and its output is binary, it is mainly used in pattern classification. BP network is multilayer feedforward neural network, and its transfer function is s-type function. Therefore, its outputs are continuous quantities from 0 to 1, and it can realize arbitrary nonlinear mapping from input to output. Because the learning algorithm of back propagation is adopted to adjust weight, it is often called BP network. After determining BP network structure, we can train it with the sample set of output and input, and adjust and learn the weight and threshold of network so that we can make network realize given mapping relation of input and output. Trained BP network can also be given proper output for input that sample set does not include, while the property can be called generalization function. In view of function fitting, this shows that network has interpolation function. At present, in the practical application of artificial neutral network, most of neutral network formulas adopt BP network and its changed forms.

After determining BP network structure, we can train network through input and output sample set. That is, we learn and adjust weight and threshold of network, so that we can realize the mapping relation of given output and input. BP network is mainly used in following occasions:

1. Function approximation: to train one network with the usage of the input vector and correspondent output vector to approximate one function.
2. Pattern recognition: to make connections between specific output vector and correspond input vector.
3. Classification: input vectors are classified in defined proper manners.
4. Data compression: to decrease the dimension of output vectors to be convenient to transmission and storage.

7.1.4.1 BP Neuron Model

One BP neuron model with R inputs is shown in Figure 7.7.

Where every input is connected by the next layer through proper weight, and the output of network can be expressed as $a = f(w^*p + b)$.

Figure 7.7 BP neuron model.

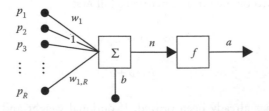

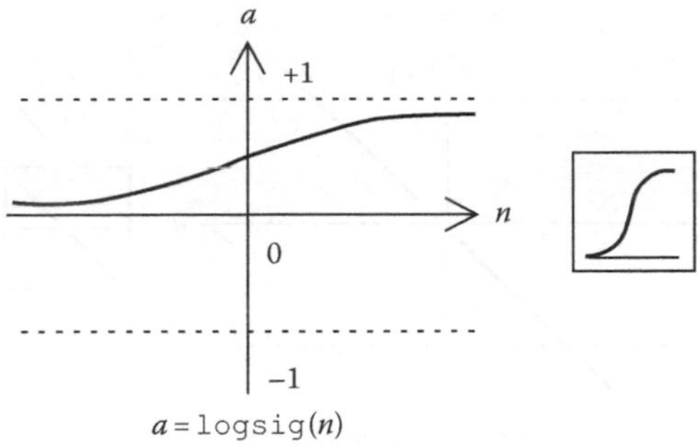

Figure 7.8 Log-sigmoid transfer function.

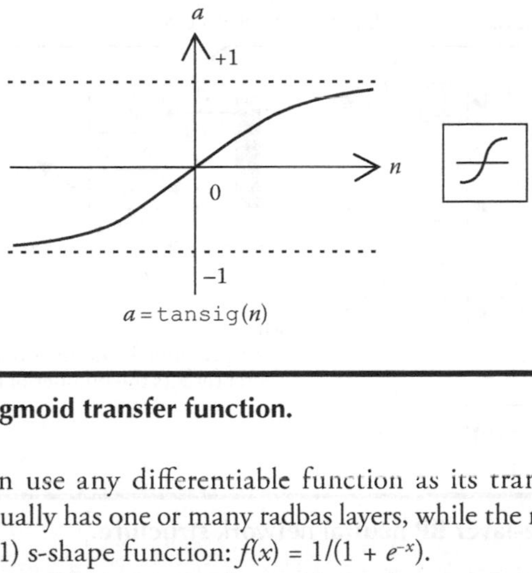

Figure 7.9 Tan-sigmoid transfer function.

BP neuron can use any differentiable function as its transfer function. BP neutral network usually has one or many radbas layers, while the radbas layer neuron usually adapts (0, 1) s-shape function: $f(x) = 1/(1 + e^{-x})$.

There are common functions such as s-shape function log-sigmoid shown in Figure 7.8, s-shape tan-sigmoid shown in Figure 7.9 and purelin function shown in Figure 7.10, and the neuron of output layer often adapt purelin, and the sort of multilayer neutral network can learn the nonlinear relation of output and input.

7.1.4.2 Network Structure of BP Neutral Network

Figure 7.11 depicts one neuron with R inputs, s-log-sigmoid transfer function, and one radbas BP network. If the last layer of BP network is sigma-shaped neuron, then the output of whole network will be limited between 0 and 1; if the last layer of BP network is linear neuron, then the outputs of whole network can be an arbitrary value.

BP neutral network can also possess multilayer structure, and their middle parts, apart from the input and output layers, are called the radbas layer. These

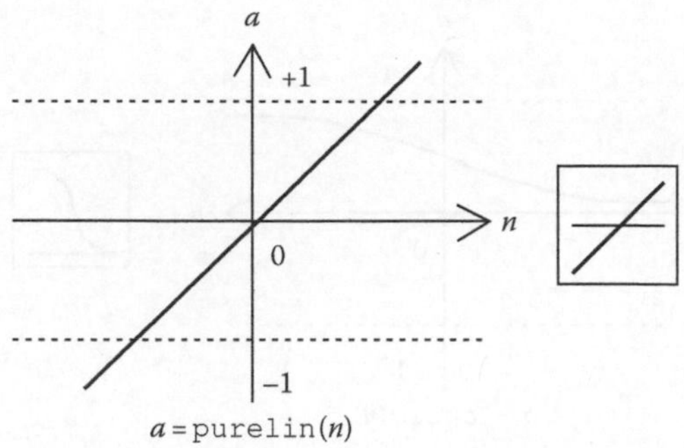

Figure 7.10 Linear transfer function.

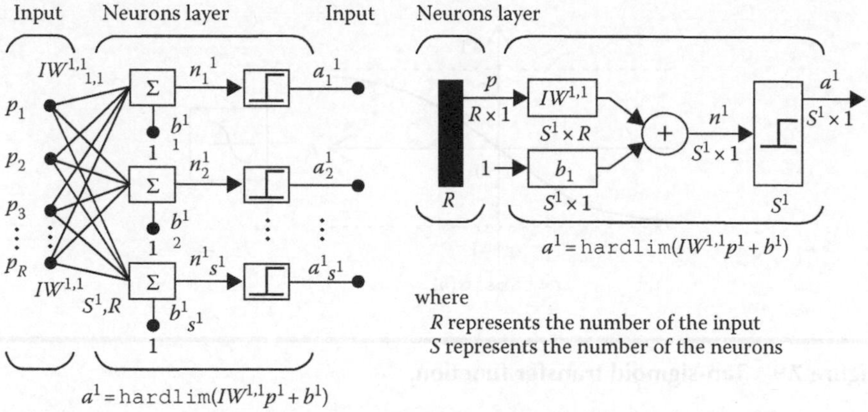

Figure 7.11 Single-layer BP neutral network structure.

radbas layers often adopt s-shape neuron, while output layers adopt linear neuron. The multilayer neutral network can learn the nonlinear relation between output and input, and the linear output layer ensues that neutral output is in the range between −1 and 1. One double-layer BP neutral network is shown in Figure 7.12.

The radbas layer of network adapts s-tan neuron, and the purelin layer adapts linear neuron, so the BP neutral network system can approximate any continuous function, if radbas layer covers enough neurons, then it also can approximate any discontinuous function with limited breakpoint.

7.1.4.3 BP Algorithm

BP algorithm is one of the simplest and most general methods in the multilayer neutral network of supervising training, and it is the natural prolongation of least-mean-squared (LMS). There are two basic operational patterns for BP neutral

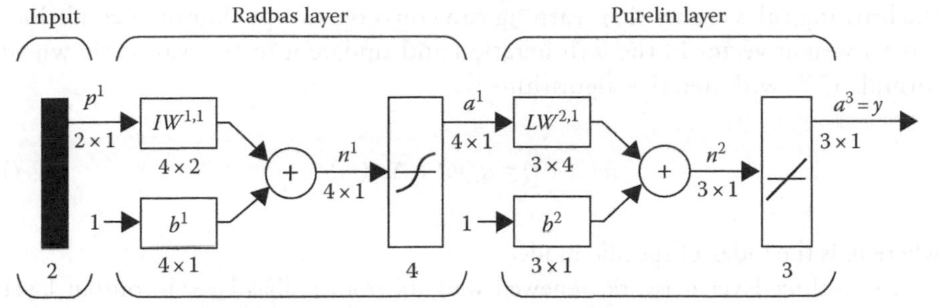

Figure 7.12 Double-layer BP neutral network structure.

network: feedforward and learning. The basic learning method of BP neutral learning is that it begins with an untrained network and supplies the input layer with a trained model, and then determines output values of the purelin layer through the network transfer signal, while any difference between the compared output value and the target value corresponds to an error. The error or criterion function is some scalar function of weight, and it will reach minimum when neutral outputs match with expected outputs. The weights are adjusted in the direction of reducing error, then the learning rules of single model BP network are given.

Trained error of single model is defined as

$$J(w) \equiv \frac{1}{2}\sum_{k=1}^{c}(t_k - z_k)^2 = \frac{1}{2}\|t - z\|^2 \tag{7.5}$$

where
- t_k is the expected output value of output terminal (given by target value)
- z_k is the actual values
- t and z are the target vector and network output vectors of length c
- w represents all weights

The learning rule of back propagation is

$$\Delta w = -\eta \frac{\partial J}{\partial w} \tag{7.6}$$

where η is learning rate and represents relative variation scale of weight.

The learning rule of back propagation is based on negative gradient declining method. First, the weights are initialized as random values, then they are adjusted in the direction of reducing errors. The advantage lies in the fact that it can reduce criterion function with only one step in the space of weights, and the learning rules ensues that learning can converge except abnormal condition. Take a weight vector in the mth iteration and update it in the way as shown in formula (7.7) with iterative algorithm

$$w(m+1) = w(m) + \Delta w(m) \tag{7.7}$$

where m is the index of specific mode.

As to three-layer network, renewed weights from radbas layer to output layer or learning rules, and weights learning rules from input layer to radbas layer are introduced, respectively, as follows:

Renewed weights from radbas layer to purelin layer or learning rules

$$\Delta w_{kj} = \eta(t_k - z_k) f'(net_k) y_j \tag{7.8}$$

where y_j is the output of radbas layer unit, $net_k = w_k^t y$ $z_k = f(net_k)$.

Learning rules from input layer to radbas layer

$$\Delta w_{ji} = \eta \left[\sum_{k=1}^{c} w_{kj} \delta_k \right] f'(net_k) x_i \tag{7.9}$$

where $\delta_k = (t_k - z_k) f'(net_k)$.

7.1.5 Radial Basis Networks

BP network is the core part of feedforward neural network, which embodies the essence part of ANN. When it is, however, used in function approximation, adjusting weights adopt negative gradient declining method. But there exists limitations of adjusting weights such as the slow speed in convergence and local minimum. Radial basis network is a kind of network that is superior to BP networks in the capability of approximation, the ability of classification, learning speed, and so on.

7.1.5.1 Radial Basis Neurons Model

One PDF neurons model with R inputs is shown in Figure 7.13.

Where the transfer function of radial basis neurons is $radbas(n) = e^{-n^2}$, as shown in Figure 7.14.

The biggest output value of transfer function of radial basis is 1, and the closer the vector distance between input value and weight is, the bigger the output value is.

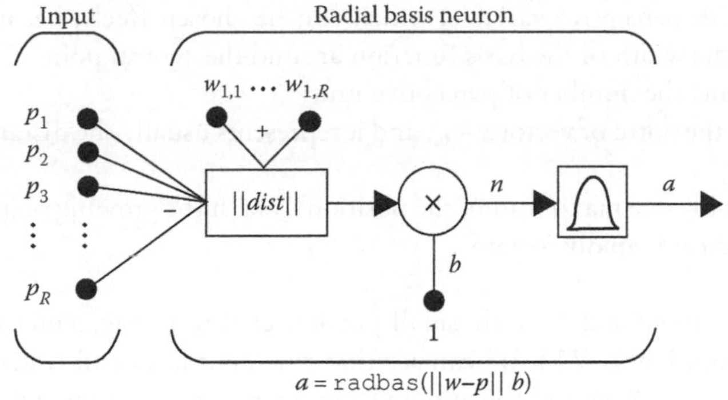

Figure 7.13 Radial basis neurons with *R* inputs.

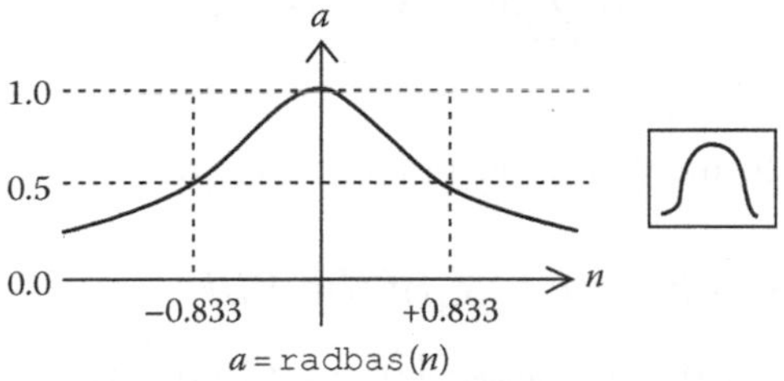

Figure 7.14 Transfer function of radial basis.

The forms of the RBF are as follows:

$$f(x) = \exp^{-(x/\sigma)^2} \tag{7.10}$$

$$f(x) = \frac{1}{(\sigma^2 + x^2)^\alpha} \quad \alpha > 0 \tag{7.11}$$

$$f(x) = (\sigma^2 + x^2)^\beta \quad \alpha < \beta < 1 \tag{7.12}$$

All the above functions are radial symmetry, and the most commonly used RBF is Gaussian function shown in formula (7.13):

$$R_i(x) = \exp\left[-\frac{\|x - c_i\|^2}{2\sigma_i^2}\right] \tag{7.13}$$

where

$i = 1,2,3,\ldots,m$

x is the input vector of n dimension

c_i is the center of the ith basis function, and its vector has the same dimension as that of x

σ_i is the ith perceptive variable, which can be chosen freely, because it determines the width of the basis function around the central point
m represents the number of perceptive unit
$||x - c_i||$ is the norm of vector $x - c_i$, and it represents usually the distance between x and c_i
$R_i(x)$ has only one maximum in the position c_i, with the growing of $||x - c_i||$, $R_i(x)$ will decrease rapidly to zero

For the given input $x \in R^n$, only small portion closing to the center of x will be activated. From Figure 7.13, we can see that the input layer can realize nonlinear mapping from x to $R_i(x)$, and the purelin layer can realize linear mapping from $R_i(x)$ to y_k, that is,

$$y_i = \sum_{i=1}^{m} w_{ik} R_i(x) \quad k = 1,2,\ldots,p$$

where p represents the number of the purelin layer node.

7.1.5.2 The Network Structure of the RBF

Radial basis neural network is consisted of three layers, which are input layer, radbas layer, and purelin layer, and the input node only transfers the input information to radbas layer, while radbas node is composed of radio pattern function like Gaussian function, and the purelin node usually is simple linear function. The constructions can be seen in Figure 7.15.

Radial basis neural network is consisted of two neurons: radbas layer and purelin layer, and its structure can be seen in Figure 7.16.

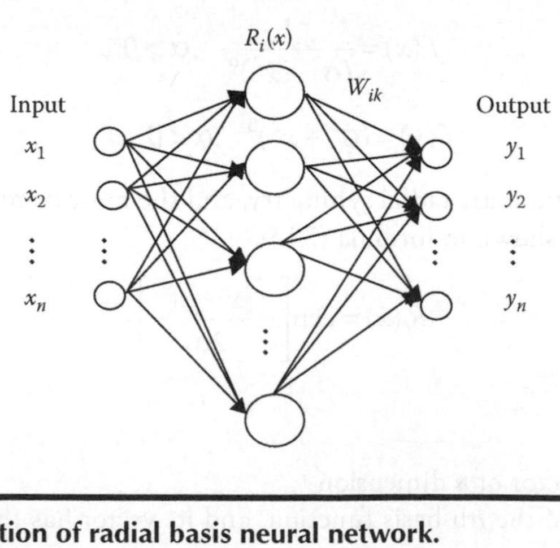

Figure 7.15 Function of radial basis neural network.

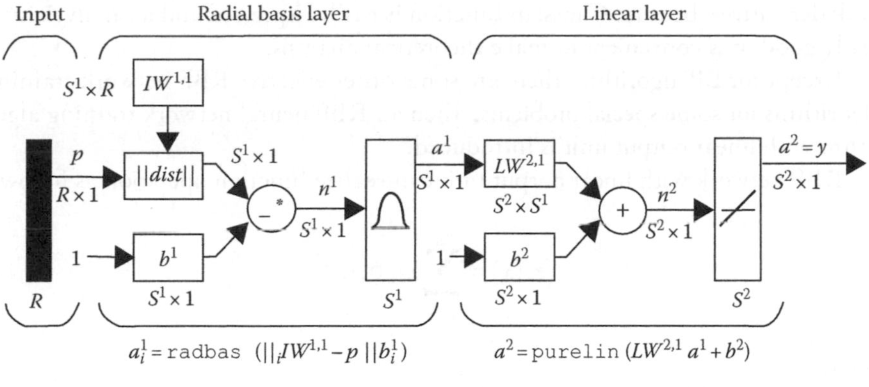

Figure 7.16 Structure of radial basis neural network.

In Figure 7.16, R represents the dimension of network input, and S^1 represents the number of neurons in the first layer, while S^2 represents the number of neurons in the second layer, and a_i^1 represents the ith element of output a^1 in the first layer, ${}_iIW_{1,1}$ represents the ith element of weight matrix ${}_iIW_{1,1}$ in the first layer, $\|dist\|$ represents the Euclidean distance between input p and weight matrix ${}_iIW_{1,1}$, and the symbol "." represents the relation of multiplication between elements in output of $\|dist\|$ and threshold value b^1.

The basis function in radbas layer node will produce partial respondence for the input signal. That is to say, when the input signal is getting close to the center of the basis function, the radbas layer node will produce stronger output. From this point of view, we can come to a conclusion that the network has local approximation capability; therefore, RBF network is also known as regional perceptive network.

7.1.5.3 Realization of the Algorithm of RBF Neural Network

The learning revision of RBF connecting weights in neural network can still adopt BP algorithm. Because $R_i(x)$ is a Gaussian function, for any x it will exists $R_i(x) > 0$, so that it will lose the advantage of partial weight adjustment, while when x is getting away from the basis function center c_i. $R_i(x)$ can become small enough and be treated as 0. Therefore, only when $R_i(x)$ is bigger than a certain value (such as 0.05), the relevant weight w_{ik} will be revised. After dealing with in this way, RBF neural network possesses the advantage that learning convergence of the partial approximation network is fast. Meanwhile, this approximate treatment can overcome the disadvantages that Gaussian function does not possess tightness.

The adoption of Gaussian function possesses the following advantages: simple expression form, even if there are multivariable inputs, it will not add too much complexity; radial symmetry; and fine smoothness, that is, there exists arbitrary rank derivatives; because Gaussian function is easily expressed and its analyticity is really good, it is convenient to make theoretical analysis.

Except for BP algorithm, there are some other effective RBF network training algorithms for some special problems. Then an RBF neural network training algorithm with linear output unit is introduced.

RBF network with linear output unit can realize function operation as follows:

$$z_k(x) = \sum_{j=0}^{n_H} w_{kj}\phi_j(x) \tag{7.14}$$

where

n_H represents the number of radbas units
$\phi_j(x)$ represents spline function

It contains one offset unit $j = 0$ in formula (7.14). If we define vector $\mathbf{\Phi}$ whose elements are the outputs of radbas layer and a matrix W whose elements are the weights from radbas layer to purelin layer, then formula (7.14) can be written as $z(x) = W\mathbf{\Phi}$. The minimum criterion function is as follows:

$$J(w) = \frac{1}{2}\sum_{m=1}^{n}\left\| y(x^m; w) - t^m \right\|^2 \tag{7.15}$$

T represents the matrix that is formed of target vector, while $\mathbf{\Phi}$ represents the matrix whose column vectors are $\mathbf{\Phi}$, then the solutions of the weights should meet

$$\mathbf{\Phi}^t\mathbf{\Phi} W^t = \mathbf{\Phi}^t T \tag{7.16}$$

The solutions can be written directly as

$$W^t = \mathbf{\Phi}^+ T \tag{7.17}$$

where $\mathbf{\Phi}^+$ is the generalized inverse matrix $\mathbf{\Phi}$.

One of the advantages of RBF with linear output unit or RBF network is that its solution can be acquired only by standard linear technique. However, the calculation amount of inversion for large matrixes is comparatively large. Therefore, the above methods can only be restrictively used in solving problems with medium size data.

Theoretically, RBF neural network can approximate any continuous nonlinear function like BP neural network. The main differences between RBF and BP lie in the fact that action functions are different, and the radbas node of BP neural network adopts sigmoid function, whose value is not zero within the infinite range of input universe, while the action function of BRF neural network is partial.

7.1.6 Probabilistic Neural Network

Probabilistic neural network (PNN) is a kind of radial basis neural network that is suitable for classification pattern, and it also enjoys good reputation in the classification systems. PNN holds very high standards for storage space, especially, when the training sampling size is very large. The main advantages of PNN are that the learning speed is fast and their design is direct without depending on training. Only once training can ensure decision interface to get closer to the optimal Bayesian decision boundary with adding training units, and the newly training sample can be easily added to training classifier; therefore, this feature is significant in online application. Furthermore, the generalization of the PNN is strong.

7.1.6.1 PNN Structure

PNN structure is shown in Figure 7.17.

As shown in this figure, the input layer of PNN is consisted of q input neurons, and each input neuron is connected with n radbas layer neurons, while each radbas layer neuron is connected with one of k purelin layer neurons. The revisable weight coefficient is represented by the connection line from input layer to radbas layer, and these weight coefficients can be achieved by training and expressed by vector w. Every purelin layer category neuron will calculate the sum of the output of radbas layer units connected with it.

PNN is consisted of a radial basis neural network layer and a competition layer, and the system structure can be seen in Figure 7.18. When an input is provided, the first layer can calculate the distance between input layer vector and the input trained vector, then a new vector is generated, and vector elements represent the degree of closeness between the two vectors. The second layer can generate a probability vector by taking the sum of the input contribution as its net output of the corresponding classification input. The third layer is able to choose the maximum

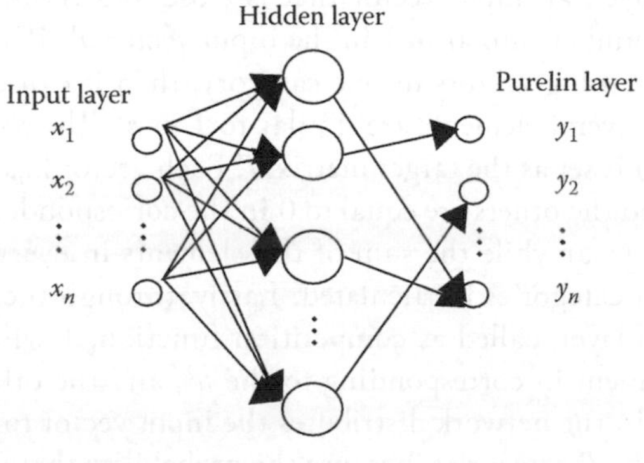

Figure 7.17 A probabilistic neural network.

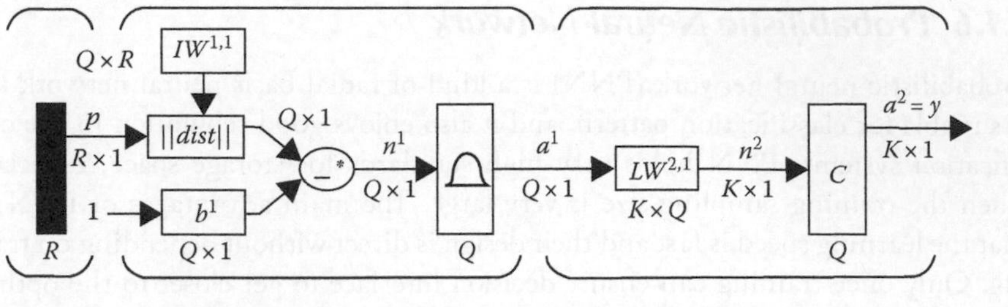

Figure 7.18 PNN system structure.

probability value through a competition conversion function generated by the output of the second layer, and then make the maximum category be 1, correspondingly, the other categories be 0.

In Figure 7.18, $a_i^1 = radbas\left(\|{}_iIW_{1,1} - p\|b_i1\right)$, $a^2 = compet(LW_{2,1}a^1)$

R is the element number in input vector.
a_i^1 is the ith element in the first layer of output a^1.
${}_iIW_{1,1}$ is a vector consisted of the units in ith line of $IW_{1,1}$, which are the elements in the ith line of the first layer of weight matrix $IW_{1,1}$.
Q is the number of neurons in the first layer (input/target logarithm), that is, the number of training samples.
K is the number of neurons in the second layer, that is, the number of classification.

Suppose there exists Q input vector/target vector, and each target vector contains K elements, where one element is 1, while the others are 0. Thus, every input vector is connected with one of the K categories. Input weight $IW_{1,1}$(*net.IW*{1, 1}) of the first layer is set as matrix transposition P^t consisted of training samples. When an input sample is provided, $\|dist\|$ will generate a vector, and its vector element represents the closeness between the input and the training vector set. Through radbas layer, an input vector that is close to a training vector will be represented by a number similar to 1 in the input vector a^1. If an input is getting closer to several training vectors in one category, then it can be represented by the numbers that several elements are similar to 1 in a^1. The second layer weight $IW_{2,1}$(*net.IW*{2, 1}) is set as the target matrix T. Each vector has only one element that equals to 1 and the others are equal to 0 in the correspondent line designated as input category, meanwhile the sum of the elements in every a^1 category that belongs to K input categories is calculated. Lastly, through the conversion function of the second layer, called as competition function, 1 will be generated by the maximum element in corresponding to the n^2, and the others are generated as 0. In view of this, the network distributes the input vector to the category that elements are 1 of the K categories, because the probability that they belong to the category is higher.

7.1.6.2 Realization of PNN Algorithm

First of all, each sample x in training sample set is normalized as unit length, that is to say, $\sum_{i=1}^{q} x_i^2 = 1$. The first sample, which is normalized, is put beyond the input neural layer. At the same time, the connection between the input neuron and the first radbas neuron is regarded as $w_1 = x_1$.

Then, a connection between the first neuron of the radbas layer and the neuron that represents x_1 in purelin layer is established. The remaining residual neurons in radbas layers are treated with same repetitive process, that is, $w_k = x_k$, where $k = 1,2,\ldots,n$. Finally, such network is obtained: the neurons of the input layer are fully connected with those of radbas layer, while the connection between the neurons from radbas layer to purelin layer is sparse.

The trained network can realize classification in the following way: first of all, one normalized test sample x is provided to input node, next the inner product of the each radbas neuron is calculated, so that we can obtain a net activation shown in formula (7.18),

$$net_k = w_k^t x \tag{7.18}$$

And then a net_k can be obtained. The results between every neuron of the purelin layer and all neurons connected in the radbas layer are added. Nonlinear function is $e^{(net_k-1)/\sigma^2}$, where σ is a parameter set by user, which represents the width of valid Gaussian window.

1. PNN training algorithm

   PNN training algorithm is as follows:

```
begin
j = 0, a_ji = 0
while j<>n do
  begin
    j = j + 1
    x_jm = x_jm / (Σ_i^q x_ji^2)^(1/2)
    w_jm = x_jm
      if x ∈ w_i then a_ji = 1
    end
end
```

   where $j = 1,2,\ldots,n$, $m = 1,2\ldots,n$, $i = 1,2,\ldots, k$, x_{jm} represents the mth component in jth sample, and x_{jm} represents the weight coefficient that connects neurons from the mth component to the jth radbas neurons.

2. PNN classification algorithm
   The training network can realize classification with the following algorithm:

```
   begin
   m = 0, x ← test points
 while m<>n do
   begin
 m = m + 1
 net_m ← w_m^t x
 if a_mi = 1 then g_i = g_i + exp (net_m - 1)/σ²
 end
 class=arg max_i g_i(x)
     return class
 end.
```

7.2 Knowledge Discovery in Databases Based on the Hybrid of VPRS and Neural Network

The process of knowledge discovery in databases contains derivation, explanation, and representing knowledge. The data in databases is often incomplete, inconsistent, unreliable, or inaccurate, and this will lead to uncertainty of knowledge which may also be resulted from the adoption of ambiguous words. Therefore, it is necessary for the tools of the knowledge discovery in databases to be able to deal with uncertain data. To do with the complexity of the real world, an ideal knowledge discovery system should be equipped with the following features:

1. It can effectively deal with the incomplete data, and the accuracy caused by incomplete data will be expressed clearly in rule strength.
2. It can select background knowledge according to the fact whether the knowledge exists or not. That is to say, on the one hand, background knowledge can be used flexibly during the process of knowledge discovery; on the other hand, if it cannot obtain background knowledge, it still can work.
3. Deviation can be selected and adjusted flexibly, and be applied in restricting and searching for control.
4. The variation of data can be treated easily. Because data in most databases frequently changes, such as addition, deletion, or updation, one good solution must be able to conveniently deal with the changing data for actual application.
5. The process of knowledge discovery can be conducted in a distributed cooperation mode.

Obviously, there is no method that can contain all the above features, while the only solution is to build a heterozygous method to combine some technologies together. Rough set theory (RST) is a strong data reasoning tool used in knowledge discovery, and attribute can be approximated in the rough set data analysis, implying that when one object is differentiated from other objects, some redundant attributes without any effect can be deleted on the condition that there is no information lost. In recent years, some experts have explored many methods of the learning classification rules based on the rough set, and these methods are decision in essence, that is to say, only when the data is consistent with those stocked in the databases, the learning classification rules can be deduced. The decision features are suitable for some application, but when there is some interference such as noise, classic rough set as a general classification method still has following problems: the over-fitting of data will reduce its prediction ability for new object; there exists the inefficiency to deal with noise, for example, 99 percent consistency and 1 percent inconsistency with databases will result from noise data, and this problem will prevent us from finding some valuable classification knowledge. The VPRS model can consider the uncertain relationship and identify strong rules by introducing a confidence threshold value β, so that it possesses error-tolerance. The final result realized by VPRS is to predict new collected data with rules generated. Neural network is dependent on the training data for system programming, and when the rules are unknown or the rules are too long to be deduced, neural network can identify the potential knowledge quickly by analyzing cases with historical data automatically. The meaning of neural network lies in pattern recognition, learning, classification, generalization, and input with incomplete or noise data. When the rules are unknown, neural network can complete amendment by renewing to training with updated dataset, so that it will exclude program modification and rule construction. After network is trained, neural network also possesses a fast running feature. So it can provide some methods to do with difficult problems. Neural network can make up for the demerits of VPRS and eventually generate full prediction formed by the new cases with data. But neural network cannot provide any detailed explanation for how the input attributes are deduced, the best way of expressing the input data, the selecting of network system structure, the number of neurons and layers, and so on. The limitation of neural network is not able to determine the relative significance of input data, so it is necessary to filter useful information from the massive input with noise. In addition, once neural network is put into practical use, it will be expensive to collect unnecessary data, even if the unnecessary data may interrupt neural network of making explanation for the final prediction. The complementarity between neural network and VPRS allows the combination of them to make the tool that possesses much stronger function than they posses individually. In the research of this chapter, VPRS is used as a pretreatment data tool for neural network to extract probability decision rules from information system. Then we provide dataset approximated by β as neural network input data, while neural network model chooses BP, RDF, and probability neural network to eventually generate full prediction for the new cases. Taking VPRS model instead of SPRS model as pretreatment model, through setting confidence threshold value β, can avoid the situation of being not able to generate rules due to several possible supernormal objects, such as the percentage of contradictory object below 1 percent. If β is set at 0.99, we can ignore the supernormal objects as noise and generate rules from them. Users can control the generation of models by setting the two parameters: confidence threshold value β and category quality threshold value $\gamma^{\beta}(P, D)$. Confidence threshold value can change detail value in rules. While classification quality threshold value is used to change the probability, which meets the object, it is useful to change approximate set of VPRS to understand the mode in data. Generally speaking, one low confidence threshold value β and one classification quality threshold value identify general and uncertain modes in data. If higher mode is identified and more precise knowledge is acquired, confidence threshold value should be set higher and classification quality threshold value β should be set lower to identify weaker mode. The process of knowledge discovery and prediction based on the hybrid between VPRS and neural network includes the following two stages: (1) the reduction of information system and knowledge discovery with the application of VPRS; (2) the classification and prediction based on training network of β-reduct dataset.

The process of the reduction of information system and knowledge discovery with the application of VPRS is consisted of the following: collection, selection, and pretreatment of the data; construction of decision table; searching of β-reduct; and generation of probability decision rules.

7.2.1 Collection, Selection, and Pretreatment of the Data

The first step is the collection and selection of data. In this step, we should take consideration of usable and reliable data that should be accepted in the form of matrix for training case set, where line represents examples or objects, column represents attribute. Every case is made up of input variable value and output variable value. Then we will pretreat these data to construct decision table, while pretreatment includes discretization of continuous data, adjustment of the defined attribute value, removal of outside value, management of missing value, and so on. The problem of data missing can be solved through providing a default value for every important attribute. For example, we can regard the past data as a default value, or delete the missing default value.

7.2.2 Construction of Decision Table

The original data in the decision table can be described easily, while the input constant is defined as condition attribute set (C), the output constant is defined as decision attribute set (D), and the set of cases is defined as universe (U). Every interval attribute $p \in C$ is converted into words or digits appointed by value universe

V_p according to its value, such as high, middle, low, or 1, 2, 3, and so on. Every defined attribute $q \in P$ can be divided into the sub-attribute value universe (V_{qi}), $(i = 1,2,\ldots,n)$, $A = C \cup D$, $C \cap D = \emptyset$, f: $U \times A \rightarrow V$, then a decision table $S = (U, C, D, V, f)$ is constructed. The decision table is the knowledge representation system in VPRS model.

7.2.3 Searching of β-Reduct and Generation of Probability Decision Rules

The aim of this step is to find the minimum related attribute subset that can maintain its original classification of the decision table for the given confidence threshold value β, and it is a method for the generation of rules and identification of the most meaningful and important variables in neural network. It will result in the fact that we can achieve success in the management of system structure of the neural network. And also, it can provide a reasonable explanation for the problem. Many reducts can be found in the stage of analyzing the decision table, and some experts have regarded those most frequently occurred and most stable reduct as the best reduct. The reducts of condition attribute set in the initial sample decision table with highest occurrence frequency is called dynamic reduct, and it has already been applied to the system ROSETTA. This chapter adopts the standard of the most suitable reducts equipped with a minimum attribute reducts, if there exists two or more reducts. In addition, the probabilistic decision rules will be formed according to their approximation, which is with the minimum combined attribute number when the same quantity of attributes are facing with the same or more approximates. The final result of VPRS is that the extracted rules from one decision table are used in predictive classification of new objects, and there exists four conditions that new objects match with rules deduced from themselves:

1. New object matches with one rule.
2. New object matches with more than one rule deduced, but its decision class is the same.
3. New object can match with several rules in different decision classes.
4. New object cannot match with any rules.

Under the circumstances (1) and (2), the result is unique, and it can give directly classification or decision suggestion according to the rules deduced by VPRS model; while it cannot give prediction directly according to the generated rules under the circumstances (3) and (4).

Taking β-reduct in the first stage as the input pattern of neural network to train network, make classification and prediction; then generate system knowledge base. The learning process includes the following: to accept reduced training sample, to complete the training process through any learning algorithm of neural network, and to predict for new object at last. The flow chart of

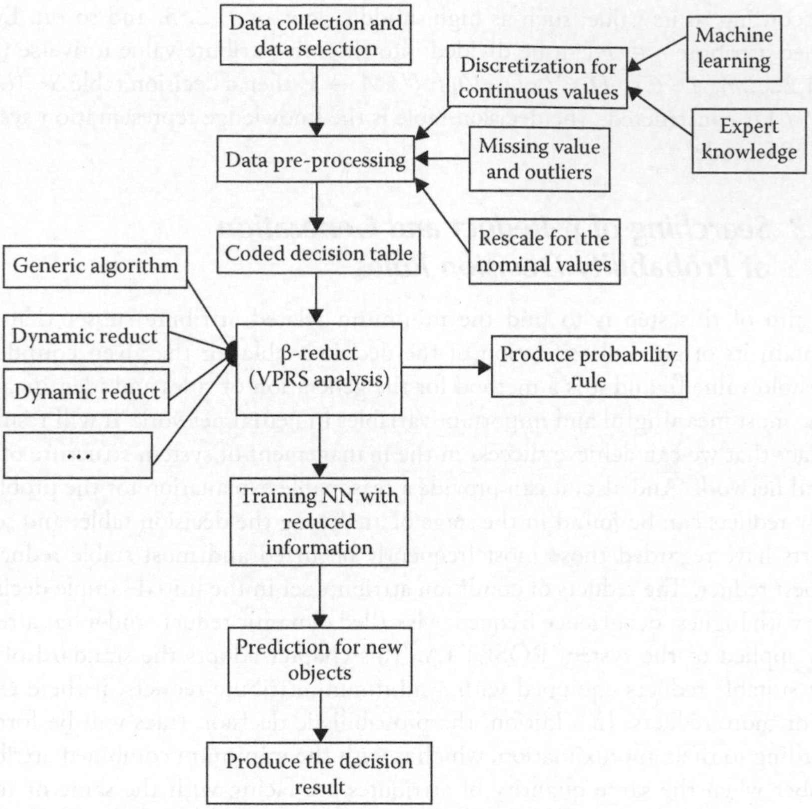

Figure 7.19 Procedure of the hybrid approach of VPRS and neural network.

the knowledge discovery approach of the hybrid between neural network and VPRS is illustrated in Figure 7.19.

> **Example 7.1** Take the dataset that can predict failure companies, for example, to illustrate the procedure of the methods mentioned above. The dataset is composed of manufacturing companies selected arbitrarily from the FAME database, in which 30 manufacturing companies are non-failure companies and the others are failure companies. Every company includes 12 condition attributes: c_1(sales), c_2(ROCE), c_3(FFTL), c_4(GEAR), c_5(CLTA), c_6(CACL), c_7(QACL), c_8(WCTA), c_9(LAG), c_{10}(AGE), c_{11}(CHAUD), c_{12}(BIG6), and a decision attribute d, while "1" represents non-failure company and "2" represents failure company. Select 40 firms randomly from the dataset, of which 20 are failure companies and 20 are non-failure companies, then take them as training sample set and the remaining 20 companies as simulation sample set. Discrete the consecutive condition attribute with FULINTER discrete method, and the discrete intervals are shown in Table 7.1, while the training sample sets and simulation sample sets are shown in Tables 7.2 and 7.3.

Table 7.1 Discrete Intervals of Conditional Attributes

Attributes	*Interval "0"*	*Interval "1"*	*Interval "2"*	*Interval "3"*
c_1	[2857.00, 5694.00]	[5694.00, 32683.50]	[32683.50, 167370]	
c_2	[–37.34970, –9.31125]	[–9.31125, –6.30660]	[–6.30660, 1.71560)	[1.71560, 33.84510]
c_3	[–0.32830, 0.02900]	[0.02900, 0.12095]	[0.12095, 0.21375)	[0.213750, 0.63120]
c_4	[0.12120, 0.57495]	[0.57495, 0.79300]	[0.79300, 1.09850)	[1.09850, 3.53360]
c_5	[0.12120, 0.46500]	[0.46500, 0.70260)	[0.70260, 1.48650]	
c_6	[0.49740, 1.16945]	[1.16945, 1.37075)	[1.37075, 4.44650]	
c_7	[0.28470, 0.96165]	[0.96165, 3.36350]		
c_8	[–0.74700, 0.05720]	[0.05720, 0.11970)	[0.11970, 0.44820]	
c_9	[55.00, 288.00]	[288.00, 421.00]		
c_{10}	[2.0, 24.5)	[24.5, 90.0]		

Table 7.2 Training Sample Sets

Attribute *Company*	c_1	c_2	c_3	c_4	c_5	c_6	c_7	c_8	c_9	c_{10}	c_{11}	c_{12}	d
n_1	0	3	3	0	0	1	0	1	0	1	0	1	1
n_2	0	3	3	1	1	2	0	2	0	0	0	0	1
n_3	1	3	3	1	2	0	0	0	0	1	0	1	1
n_4	1	0	0	3	1	1	1	2	0	0	0	1	1
n_5	1	3	3	0	1	1	0	2	0	1	0	0	1
n_6	2	3	3	2	0	2	1	2	0	0	0	1	1
n_7	1	3	2	0	0	2	1	2	0	1	0	0	1
n_8	1	1	0	2	0	2	1	2	1	1	0	1	1
n_9	0	1	0	0	1	2	1	2	1	1	0	1	1
n_{10}	0	2	0	1	0	1	0	1	1	1	1	1	1
n_{11}	0	3	2	1	1	0	0	0	0	0	0	0	1
n_{12}	1	3	3	0	0	2	1	2	0	1	0	0	1
n_{13}	2	3	3	1	2	0	0	1	1	1	1	1	1
n_{14}	0	3	3	0	0	2	1	2	0	1	0	0	1
n_{15}	2	1	0	1	0	2	1	1	0	0	1	1	1
n_{16}	1	3	2	1	1	0	0	1	0	0	0	1	1
n_{17}	1	3	3	0	0	2	1	2	0	1	0	1	1
n_{18}	0	0	0	3	2	0	0	0	0	0	0	1	1
n_{19}	1	3	2	1	1	0	0	0	0	0	0	0	1
n_{20}	2	1	0	1	2	0	0	0	0	1	0	1	1
n_{21}	0	0	0	3	2	0	0	0	1	0	0	0	2
n_{22}	0	3	3	0	0	2	1	2	0	0	0	1	2
n_{23}	2	2	2	2	2	0	0	0	0	0	0	1	2
n_{24}	1	2	2	2	2	0	0	0	1	1	0	0	2
n_{25}	1	3	3	1	1	0	0	1	0	0	0	0	2
n_{26}	1	2	2	2	1	0	0	0	1	0	0	0	2

Table 7.2 (continued) Training Sample Sets

Attribute *Company*	c_1	c_2	c_3	c_4	c_5	c_6	c_7	c_8	c_9	c_{10}	c_{11}	c_{12}	d
n_{27}	1	3	3	2	0	2	0	2	1	1	1	1	2
n_{28}	2	3	3	2	2	1	0	2	1	1	0	0	2
n_{29}	1	2	2	2	1	0	0	0	1	0	1	1	2
n_{30}	0	2	2	1	1	0	0	0	1	0	1	0	2
n_{31}	0	0	0	2	1	0	0	0	1	0	0	0	2
n_{32}	0	2	2	2	2	0	0	0	1	0	1	0	2
n_{33}	0	0	0	1	1	0	0	0	1	0	0	1	2
n_{34}	1	2	2	1	1	0	0	0	0	1	0	0	2
n_{35}	0	2	2	1	1	0	0	0	0	1	0	0	2
n_{36}	0	2	2	2	2	1	0	2	0	0	1	1	2
n_{37}	0	2	2	1	1	1	1	2	1	0	0	0	2
n_{38}	1	2	2	2	2	0	0	0	0	0	1	1	2
n_{39}	0	0	0	1	2	0	0	1	0	0	1	1	2
n_{40}	1	2	2	2	2	0	0	0	1	0	0	0	2

7.2.3.1 Searching of β-Reduct

Simulate with the dataset in Table 7.2. To identify the stronger attributes and patterns in data, confidence threshold β will be set as a higher value. Take β as 0.95 and search reducts of the dataset in decision Table 7.2 with inherit algorithm, we can get a minimum attribute β-reduct, that is, $\{c_1, c_2, c_4, c_8, c_{12}\}$.

7.2.3.2 Learning and Simulation of the Neural Network

Then construct BP neural network, RBF neural network, and probability neural network with the same dataset of β-reduction, and we design neural network based on Matlab6.0 neural network toolbox in this book.

During the design of BP neural network, function `newff(·)` is used to build a network, while the radbas layer adopts tansig neuron, and trainscg algorithm is used to train BP neural network. Net.trainparam.show presents the time of each state,

Table 7.3 Simulation Sample Sets

Attribute *Company*	c_1	c_2	c_3	c_4	c_5	c_6	c_7	c_8	c_9	c_{10}	c_{11}	c_{12}	d
n_{41}	1	3	2	1	1	2	0	2	0	1	0	0	1
n_{42}	1	2	0	3	2	0	0	0	0	1	0	1	1
n_{43}	1	3	3	0	0	2	1	2	0	1	0	1	1
n_{44}	1	3	3	0	1	2	1	2	0	1	0	0	1
n_{45}	2	3	3	0	0	2	1	2	1	1	0	1	1
n_{46}	0	2	0	0	0	2	1	2	1	0	0	1	1
n_{47}	1	3	3	1	1	1	1	2	0	0	0	1	1
n_{48}	2	3	3	0	0	1	1	1	0	1	0	1	1
n_{49}	0	1	1	1	1	1	0	2	1	1	0	0	1
n_{50}	2	3	2	0	0	1	1	0	0	0	1	1	1
n_{51}	0	3	2	1	2	0	0	1	0	0	1	0	2
n_{52}	1	0	0	2	2	0	0	0	0	0	0	1	2
n_{53}	1	0	0	2	2	0	0	0	1	1	0	0	2
n_{54}	1	3	2	2	2	0	0	0	1	0	0	1	2
n_{55}	0	2	1	1	1	0	0	0	0	0	1	0	2
n_{56}	0	2	1	2	2	0	0	0	1	1	0	0	2
n_{57}	0	0	0	3	2	0	0	0	1	0	0	0	2
n_{58}	0	3	3	1	1	1	1	2	1	0	0	1	2
n_{59}	1	2	2	1	1	0	0	0	1	1	0	0	2
n_{60}	1	2	2	1	1	1	0	1	0	1	0	0	2

while net.trainparam.epochs is the maximum training times, and net.trainparam.goal is the error threshold and the radbas layer is 5. In the end, simulate and imitate with the sim function. The program segments of building a two-layer BP neural network are illustrated as follows:

```
% Program segments of BP neural network design
load reduct.m    % Load the condition attribute set of
  β-reduct training sample
```

```
load traindatad.m    % Load decision attribute set
load reductvalidate.m    % Load condition attribute set of
  β-reduct simulation sample
p1=reductvalidate
p1=p1'
p=reduct;
b=traindatad
p=p'
t=b';
net=newff(minmax(p),[5,1],{'tansig','purelin'},'train
  scg');    % Build network
net.trainparam.show=10    % Show the time of every state
net.trainparam.epochs=300;    % maximum training times
net.trainparam.goal=1e-5;    % error threshold value
[net,tr]=train(net,p,t)
y=sim(net,p);
y
y1=sim(net,p1);
y1
```

In the design of radial basis neural network, a radial basis neural network is built based on `newrb(·)` function. `Newrb(·)` function reduces the error of network output by increasing automatically radial basis neuron, it will not finish the network training until error meets the requirement.

```
% Program segments of BP neural network design
load reduct.m;
load train datad.m
load reduct validate.m
p=reduct;
p=p';
p1=reduct validate;
p1=p1'
b=traindatad;
t=b';
net=newrb(p,t);
y=sim(net,p);
y
y1=sim(net,p1);
y1
```

In the training process of probability neural network, we create a probability neural network with the function `newpnn(·)`, then we build a target matrix of which the right position is 1 and the others are 0 with function `ind2vec(·)`. In the end, we imitate and simulate with function `sim(·)`.

```
load reduct.m load train datad.m
  load reduct validate.m
  p=reduct;
  bc=traindatad
  p=p'
  tc=bc';
  p1=reductvalidate
  p1=p1'
  t=ind2vec(tc)
  net=newpnn(p,t);
  y1=sim(net,p);
  yc1=vec2ind(y1)
  y2=sim(net,p1);
  yc2=vec2ind(y2)
```

Taking the datasets after the treatment of β-reduct in Table 7.2 as training sample sets, and the same datasets in Table 7.3 as simulation sample sets, the classification results and simulation results based on VPRSs model, hybrid between VPRSs and BP neural network, hybrid between VPRSs and RBF, hybrid between VPRSs and probability neural network are shown in Tables 7.4 and 7.5, and the classification qualities and simulation qualities are shown in Table 7.6. The imitative and simulative figures are shown in Figures 7.20 and 7.21.

From Table 7.6, Figures 7.20, and 7.21, we can know that imitation classification quality based on the hybrids that are between VPRS and BP neutral network, between VPRS and RDF neutral network, and between VPRS and PNN, all is higher than that of VPRS. While in the simulation, the simulation classification quality based on the hybrids of variable precision and BP neutral network, and of VPRS and RBF neutral network is very poor, and lower than that of VPRS, while simulation classification quality based on the hybrid of VPRS and PNN is the highest and higher than that of VPRS. Moreover, its classification and prediction results are more visual. Because, the process of setting learning parameter for BP neutral network and RBF neutral network usually contains too many attempts while PNN only needs to train one time, and we can obtain high precision of the classification and simulation. In addition, the classification results are more visual, so we can know that PNN possesses excellent classification and generalization abilities.

7.3 System Design Methods of the Hybrid of Variable Precision Rough Fuzzy Set and Neutral Network

Fuzzy neutral network combines the advantages of fuzzy logic and neutral network, and it possesses not only the functions of neutral network such as stronger classification, memory, learning, and association, but also the functions

Table 7.4 Results of Imitation Classification

Company	*BP Neural Network*	*RBF Neural Network*	*Probability Neural Network*
n_{41}	1.0133	1	1
n_{42}	1.0373	1	1
n_{43}	0.9792	1	1
n_{44}	0.9998	1	1
n_{45}	1.0007	1	1
n_{46}	1.0066	1	1
n_{47}	1.0007	1	1
n_{48}	0.9623	1	1
n_{49}	1.0139	1	1
n_{50}	1.0064	1	1
n_{51}	1.0855	1	1
n_{52}	1.0007	1	1
n_{53}	0.9584	1	1
n_{54}	1.006	1	1
n_{55}	1.053	1	1
n_{56}	1.3106	1	1
n_{57}	0.807	1	1
n_{58}	1.0168	1	1
n_{59}	0.9493	1	1
n_{60}	1.0668	1	1
n_{41}	1.9184	2	2
n_{42}	2.0265	2	2
n_{43}	1.9506	2	2
n_{44}	2.0681	2	2
n_{45}	1.997	2	2

(continued)

Table 7.4 (continued) Results of Imitation Classification

Company	*BP Neural Network*	*RBF Neural Network*	*Probability Neural Network*
n_{46}	2.0021	2	2
n_{47}	1.9954	2	2
n_{48}	2.0161	2	2
n_{49}	2.0733	2	2
n_{50}	1.986	2	2
n_{51}	1.9954	2	2
n_{52}	1.8353	2	2
n_{53}	1.8412	2	2
n_{54}	1.9857	2	2
n_{55}	1.9967	2	2
n_{56}	1.9857	2	2
n_{57}	2.0641	2	2
n_{58}	2.0219	2	2
n_{59}	1.9354	2	2
n_{60}	1.997	2	2

such as dealing with fuzzy information, executing fuzzy reasoning and parallel computing. However, when we carry out data mining for a large-scale database, fuzzy rules will take on exponential growth with the increase of input dimension; thus, it results in extremely complicate neutral network structure. How to construct initial structure of fuzzy neutral network in the imperfect expert knowledge fields, reduce searching time and space, and improve learning efficiency of network are very important issues. The variable precision RST has very good ability in knowledge acquisition and processing, and can greatly reduce the dimension of the knowledge representation space. Hybrid of VPRS and fuzzy neutral network is done according to the following steps: first, the radbas rules of sample feature are mined with the variable precision RST; then the fuzzy neutral network model is established to make a decision, while it not only simplifies the complexity of the information processing but also improves the precision of the information processing. This chapter mainly introduces general method of the hybrid of VPRS and fuzzy neutral network.

Table 7.5 Simulation Results

Company	*BP Neural Network*	*RBF Neural Network*	*Probability Neural Network*
n_{41}	1.7452	1.4391	1
n_{42}	2.0329	2.9095	2
n_{43}	0.9997	1	1
n_{44}	0.9878	1	1
n_{45}	0.9673	1.9085	1
n_{46}	1.0677	1.4234	1
n_{47}	1.6	1.2813	1
n_{48}	0.8979	1.948	1
n_{49}	1.7321	2.2998	1
n_{50}	0.2534	2.5529	1
n_{51}	1.8917	2.2967	2
n_{52}	2.0653	2	2
n_{53}	1.2561	2.9798	2
n_{54}	2.0099	2	2
n_{55}	2.0091	2	2
n_{56}	1.2757	2.3543	1
n_{57}	1.7353	1.5468	1
n_{58}	1.9878	2	2
n_{59}	0.9243	2.7285	1
n_{60}	2.0615	2	2

7.3.1 Construction of a Variable Precision Rough Fuzzy Neutral Network

For given confidence β, first of all, reduct information system with VPRS, then construct variable precision rough fuzzy neutral network according to the decision table processed with β-reduct, while variable precision rough fuzzy neutral network is usually divided into five layers as shown in Figure 7.22. It is designed according to the working process of variable precision rough theory, and in essence it can realize the reasoning system of the VPRS with fuzzy neutral network.

Table 7.6 Comparison Classification and Prediction Results

Models	*Classification Quality*	
	Training Sample Sets (Percent)	*Simulation Sample Sets (Percent)*
Variable precision rough sets	91.7	70
Hybrid of variable precision rough sets and BP neural network	100	60
Hybrid of variable precision rough sets and radial basis neural network	100	65
Hybrid of variable precision rough sets and probability neural network	100	80

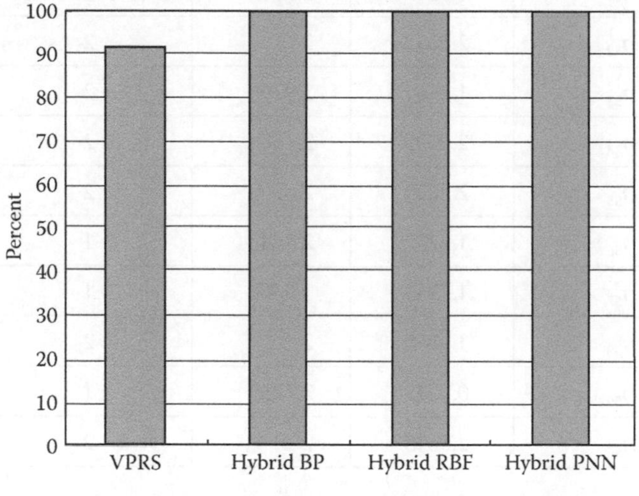

Figure 7.20 Imitation classification quality of hybrid model.

The first layer is input layer. Inputs of this layer are exact values of eigenvector $X = (x_1, x_2,\ldots, x_n)^T$, that is, they are exact condition attribute values; the number of neuron n is equal to the number of condition attribute after reduct.

The second layer is fuzzy layer. When one input vector is given to input layer, we must determine the relationship with every correspondent radbas layer.

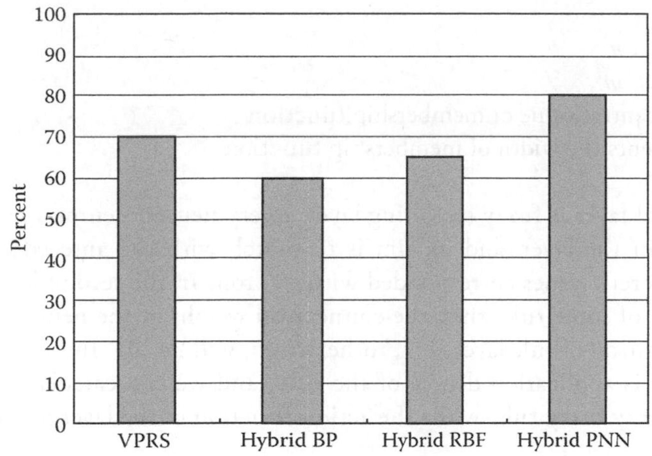

Figure 7.21 Simulation classification quality of hybrid model.

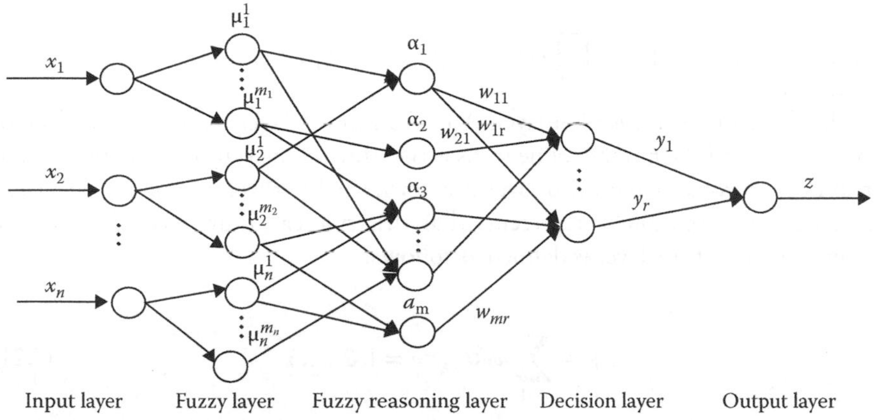

Figure 7.22 Construction of the variable precision rough fuzzy neutral network.

The function of a group of neurons connected with *r*th output node is to explain *r*th component of input vector and make exact values fuzzy. The neutral function of this layer is fuzzy membership function corresponding with discrete value. Define a fuzzy set in each discrete interval of the input component x_i, and these values are in the range between 0 and 1. The action function of neuron is a Gauss function expressed as follows:

$$\mu_i = \exp\left(\frac{-(x_i - c_{ij})^2}{\sigma_{ij}^2}\right) \tag{7.19}$$

where

$i = 1, 2, \ldots, n$

$j = 1, 2, \ldots, m_i$

c_{ij} is the central value of membership function

σ_{ij} represents the width of membership function

The third layer is fuzzy reasoning layer. Every neutral neuron represents one fuzzy rule in the layer, and its aim is to match with the antecedent of fuzzy rules. If discrete values corresponded with neurons in the second layer are rules antecedents of some rule, then the connection weight of the neuron and corresponding neuron of rule layer is "1," otherwise it will be "0," The action function of the layer is application degree of the rule, and we can calculate application degree of every fuzzy rule, while the action function of the layer neuron is given as follows:

$$\alpha_j = \min(\mu_1^{i_1}, \mu_2^{i_2}, \ldots, \mu_n^{i_n}) \tag{7.20}$$

where $j = 1, 2, \ldots, m$; $m = \prod_{i=1}^{n} m_i$; $i_1 \in \{1, 2, \ldots, m_1\}$; $i_2 \in \{1, 2, \ldots, m_2\}, \ldots, i_n \in \{1, 2, \ldots, m_n\}$.

The fourth layer is decision layer. A neuron in the layer represents the type of input object, and the number of neurons is the same as the type number of decision attribute. Neurons in the third layer are connected with corresponding decision neuron of the layer, and it represents that this rule can draw a conclusion. The action function of the layer is defined as follows:

$$y_i = \sum_{j=1}^{m} w_{ij}\alpha_j, \quad i = 1, 2, \ldots, r \tag{7.21}$$

In formula (7.21), w_{ij} represents confidence value of the rule, and the initial value is regarded as the confidence value of corresponding rules.

The fifth layer is output layer. The node number is the same as that of decision attribute, and this layer represents de-fuzzy. b_i is decision attribute value to which decisive neuron is corresponding, and it remains unchanged in the network training.

$$z = \frac{\sum_{i=1}^{r} y_i * b_i}{\sum_{i=1}^{r} y_i} \tag{7.22}$$

7.3.2 Training Algorithm of the Variable Precision Rough Fuzzy Neutral Network

Before learning of variable precision rough fuzzy neutral network, we should set the biggest training times MAXEPOCH, expected error value EG, learning rate η, and connection weight $w_{ij}(i = 1, 2,\ldots, r, j = 1, 2,\ldots, m)$ between reasoning layer and output layer.

Because fuzzy division numbers of each input component of fuzzy layer are numbers of discrete interval, the shape of membership function is given according to pretreatment results of VPRS, and it is not revised in the learning process. Neutral network of VPRS only trains connection weights $w_{ij}(i = 1, 2,\ldots, r, j = 1, 2,\ldots, m)$ between reasoning layer and output layer.

The learning algorithm of the variable precision rough fuzzy neutral network is depicted as follows:

1. Get one element in turn from a decision table and take its attribute value as input vector X.
2. Calculate membership function that each input component of X belongs to each fuzzy set according to formula (7.19).
3. Match antecedent of classification rules and calculate application degree according to formula (7.20).
4. Calculate output of each neuron of output layer according to formula (7.21).
5. Adjust connection weights according to the following decision attribute of classification rules:

$$w_{ij}(k+1) = w_{ij}(k) - \eta(y_i - d_i)\sigma_j$$

   where
   k is training times
   η is learning rate
   d_i is expected output of each neuron
6. Return to step 1 and take next element, repeat carrying out the steps from 1 to 5 until all elements are used out.
7. Calculate internet error

$$E = \frac{1}{2}\sqrt{\sum_{p=1}^{m}\sum_{i=1}^{r}(d_{pi} - y_{pi})^2}$$

   where r is the number of neurons of output layer.
8. Examine termination condition, if $E < Eg$ or $k =$ MAXEPOCH, then finish training, otherwise $k = k + 1$, turn to step 1, and begin next training.

7.4 Summary

This chapter mainly introduces a hybrid approach of VPRSs, fuzzy sets, and neural network. The aim of the hybrid is to analyze existing data and obtain pattern prediction ability from the existing patterns. VPRS is used to recognize non-decision relation from the noisy data, and it can not only obtain some important input attributes but also generate probability decision rules; take information table after reduct as the input data of neural network or construct variable precision rough fuzzy neural network and then classification and prediction can be realized with the model based on neural network. VPRS analysis adds some useful properties to neural network system, such as the compatibility with noisy data, and exclusion of redundancy and uncertainty of neural network. Attribute β-reduct avoids the over-fitting problems and reduces the data input quantity of neural network, which leads to a decrease in the number of the input nodes and weights, accelerate the process of network learning, shorten the network training time, and strengthen the explanation ability of neural network. Besides, the generated rules can be used to understand the relation between decision attribute and condition attribute. Furthermore, one advantage of VPRS is that the procedure describing the method can be run in parallel mode. The property of neural network is that of learning from examples, such as when the reduced training case is submitted to the network, a convenient part information reasoning based on data relationship discovery would be provided, and the hybrid approach can discover knowledge and predict from the fuzzy, incomplete, and noisy data, and put the structure knowledge into the system. Because database in business is always large and incomplete, the ability is extremely important for data analysis.

Chapter 8

Application Analysis of Hybrid Rough Set

In this chapter, many hybrid rough set approaches are used in transport scheme choice decision of the supply chain network, and we compare and analyze the decision results from different approaches.

8.1 A Survey of Transport Scheme Choice

Transport is the displacement of products from one place to another in supply chain. Transport cost is a crucial part of the supply chain cost, so transport plays a very important role in a supply chain. In 1996, American goods transport cost reached $455 billion, which accounted for about 6 percent of GDP. With the development of e-business and goods-to-door service, the position of transport cost in retailing is even more outstanding. Every transport decision affects profit, inventory decision, and facilities decision in supply chain, while the aim of the supply chain is not only to provide a proper response sensitivity for customers, but also to cut down on total cost. The success of a supply chain is strongly linked to the reasonable choice of transport pattern. Inventory cost is also important when transport decision is made and, if ignored, influences the operational performance of supply chains. In the process of making transport decision of the supply chain network design, enterprises have to make choices among inventory, freight, and response ability. For example, Dell's cost of packaging personal computers to customers increases, but this makes it possible to lay out all kinds of its facilities intensively. If Dell wants to cut down its transport cost, then it can either lower the sensitivity of customers' demand response or increase the number of the facilities and establish stores near its customer bases. It is important to keep a balance between transport cost and inventory cost. Quick transport pattern is suitable for high value/weight products, because decreasing the inventory is important for these products; slower transport pattern, on the other hand, is suitable for low value/weight products, because cutting down the cost is most important. The transport cost caused by supply chain has a close relationship with the response ability of the supply chain. If an enterprise reacts swiftly, and it delivers goods when receiving orders on the same day, then its delivery scale becomes smaller, which will result in high freight. If an enterprise lowers its response ability, and deliver goods intensively ordered during a long period of time, then it will gain scale economy by means of batch delivering. Delivering goods intensively when postponing the shipment and lowering the response ability can cut down the transport cost.

Ten transport schemes received by a factory manager are shown in Table 8.1. Each transport scheme includes four condition attributes $C = \{c_1,c_2,c_3,c_4\}$ and a decision attribute $D = \{d\}$, where c_1, transport cost; c_2, batch scale; c_3, inventory cost; c_4, response time; and d, the comprehensive evaluation of transport scheme.

Table 8.1 An Information System of Transport Scheme Choice Decision

Contractor	c_1	c_2	c_3	c_4	d
n_1	High	Low	High	Slow	Bad
n_2	High	High	Low	Quick	Good
n_3	Low	High	Low	Quick	Good
n_4	High	High	Medium	Quick	Bad
n_5	Low	Low	Medium	Quick	Bad
n_6	Low	Low	High	Quick	Good
n_7	Low	Low	High	Quick	Good
n_8	Low	Low	High	Quick	Bad
n_9	High	High	High	Slow	Bad
n_{10}	Low	Low	High	Quick	Good

8.2 Transport Scheme Choice Decision Undertaking No Consideration into Preference Information

8.2.1 Choice Decision Based on Rough Set

Based on rough set theory, the universe of Table 8.1 is partitioned into the following equivalence classes:

$$\frac{U}{C}=\{X_1,X_2,X_3,X_4,X_5\}$$

where $X_1 = \{n_1\}$, $X_2 = \{n_2\}$, $X_3 = \{n_3\}$, $X_4 = \{n_4\}$, $X_5 = \{n_5\}$, $X_6 = \{n_6, n_7, n_8, n_{10}\}$, $X_7 = \{n_9\}$.

$$\frac{U}{D}=\{Y_B,Y_G\}$$

where $Y_B = \{n_1, n_4, n_5, n_8, n_9\}$, $Y_G = \{n_2, n_3, n_6, n_7, n_{10}\}$.

Then the lower approximation $\underline{apr}_P(Y_B)$ of Y_B and the upper approximation $\overline{apr}_P(Y_B)$ of Y_B are, respectively,

$$\underline{apr}_P(Y_B)=\{\{n_1\},\{n_4\},\{n_5\},\{n_9\}\}$$

$$\overline{apr}_P(Y_B)=\{\{n_1\},\{n_4\},\{n_5\},\{n_9\},\{n_6,n_7,n_8,n_{10}\}\}$$

while the lower approximation $\underline{apr}_P(Y_G)$ of Y_G and the upper approximation $\overline{apr}_P(Y_G)$ of Y_G are, respectively,

$$\underline{apr}_P(Y_G)=\{\{n_2\},\{n_3\},\{n_6,n_7,n_8,n_{10}\}\}$$

$$\overline{apr}_P(Y_G)=\{\{n_2\},\{n_3\},\{n_6,n_7,n_8,n_{10}\}\}$$

Based on the genetic algorithm, the minimal reduct $\{c_1, c_3\}$ is computed, and then decision rules induced from the reduct $\{c_1, c_3\}$ are shown in Table 8.2.

8.2.2 Probability Choice Decision Based on VPRS

Based on variable precision rough set (VPRS), the universe of the Table 8.1 is portioned into the following equivalence classes:

$$\frac{U}{C}=\{X_1,X_2,X_3,X_4,X_5\}$$

Table 8.2 A Decision Rule Set Constructed for the Reduct $\{c_1, c_3\}$

Rules	*Support*	*Confidence (Percent)*
$c_1 = \text{high} \wedge c_3 = \text{high} \xrightarrow{\text{100 percent}} d = \text{bad}$	2	100
$c_1 = \text{high} \wedge c_3 = \text{low} \xrightarrow{\text{100 percent}} d = \text{good}$	1	100
$c_1 = \text{low} \wedge c_3 = \text{low} \xrightarrow{\text{100 percent}} d = \text{good}$	1	100
$c_1 = \text{high} \wedge c_3 = \text{medium} \xrightarrow{\text{100 percent}} d = \text{bad}$	1	100
$c_1 = \text{low} \wedge c_3 = \text{medium} \xrightarrow{\text{100 percent}} d = \text{bad}$	1	100

where $X_1 = \{n_1\}$, $X_2 = \{n_2\}$, $X_3 = \{n_3\}$, $X_4 = \{n_4\}$, $X_5 = \{n_5\}$, $X_6 = \{n_6, n_7, n_8, n_{10}\}$, $X_7 = \{n_9\}$.

$$\frac{U}{D} = \{Y_B, Y_G\}$$

where $Y_B = \{n_1, n_4, n_5, n_8, n_9\}$, $Y_G = \{n_2, n_3, n_6, n_7, n_{10}\}$.

If confidence threshold $\beta = 70$ percent, then the β-lower approximation $\underline{apr}_P^{\beta}(Y_B)$ of Y_B and the β-upper approximation $\overline{apr}_P^{\beta}(Y_B)$ of Y_B are, respectively,

$$\underline{apr}_P^{\beta}(Y_B) = \{\{n_1\}, \{n_4\}, \{n_5\}, \{n_9\}\}$$

$$\overline{apr}_P^{\beta}(Y_B) = \{\{n_1\}, \{n_4\}, \{n_5\}, \{n_9\}\}$$

while the β-lower approximation $\underline{apr}_P^{\beta}(Y_G)$ of Y_G, and the β-upper approximation $\overline{apr}_P^{\beta}(Y_G)$ of Y_G are, respectively,

$$\underline{apr}_P^{\beta}(Y_G) = \{\{n_2\}, \{n_3\}, \{n_6, n_7, n_8, n_{10}\}\}$$

$$\overline{apr}_P^{\beta}(Y_G) = \{\{n_2\}, \{n_3\}, \{n_6, n_7, n_8, n_{10}\}\}$$

Based on the genetic algorithm, the minimal β-reduct $\{c_1, c_3\}$ is computed, and then the probability decision rules induced from the β-reduct $\{c_1, c_3\}$ are shown in Table 8.3.

Table 8.3 A Probability Decision Rule Set Constructed by the β-Reduct $\{c_1, c_3\}$

Rules	Support	Confidence (Percent)
$c_1 = \text{high} \wedge c_3 = \text{high} \xrightarrow{\text{100 percent}} d = \text{bad}$	2	100
$c_1 = \text{high} \wedge c_3 = \text{low} \xrightarrow{\text{100 percent}} d = \text{good}$	1	100
$c_1 = \text{low} \wedge c_3 = \text{low} \xrightarrow{\text{100 percent}} d = \text{good}$	1	100
$c_1 = \text{high} \wedge c_3 = \text{medium} \xrightarrow{\text{100 percent}} d = \text{bad}$	1	100
$c_1 = \text{low} \wedge c_3 = \text{medium} \xrightarrow{\text{100 percent}} d = \text{bad}$	1	100
$c_1 = \text{low} \wedge c_3 = \text{high} \xrightarrow{\text{75 percent}} d = \text{good}$	4	75

8.2.3 Choice Decision Based on Grey Rough Set

The universe of Table 8.1 is partitioned into the following equivalence classes:

$$\frac{U}{C} = \{X_1, X_2, X_3, X_4, X_5\}$$

where $X_1 = \{n_1\}$, $X_2 = \{n_2\}$, $X_3 = \{n_3\}$, $X_4 = \{n_4\}$, $X_5 = \{n_5\}$, $X_6 = \{n_6, n_7, n_8, n_{10}\}$, $X_7 = \{n_9\}$.

$$\frac{U}{D} = \{Y_B, Y_G\}$$

where $Y_B = \{n_1, n_4, n_5, n_8, n_9\}$, $Y_G = \{n_2, n_3, n_6, n_7, n_{10}\}$.

If the grey level $g_c = 3$, then the minimal g_c-reduct $\{c_1, c_3\}$ based on the grey rough set genetic algorithm is computed, and the probability decision rules induced from the g_c-reduct $\{c_1, c_3\}$ are shown in Table 8.4.

8.2.4 Probability Choice Decision Based on the Hybrid of VPRS and Probabilistic Neural Network

Taking, respectively, the source datasets and the datasets with β-reduct as training sample sets, we can obtain classification results shown in Table 8.5 based on the hybrid of the VPRS and probabilistic neural network.

Table 8.4 Decision Rules for g_c-Reduct $\{c_1, c_3\}$

Rules	*Support*	*Grey level*
$c_1 = \text{high} \wedge c_3 = \text{high} \xrightarrow{g_c=0} d = \text{bad}$	2	$g_c = 0$
$c_1 = \text{high} \wedge c_3 = \text{low} \xrightarrow{g_c=0} d = \text{good}$	1	$g_c = 0$
$c_1 = \text{low} \wedge c_3 = \text{low} \xrightarrow{g_c=0} d = \text{good}$	1	$g_c = 0$
$c_1 = \text{high} \wedge c_3 = \text{medium} \xrightarrow{g_c=0} d = \text{bad}$	1	$g_c = 0$
$c_1 = \text{low} \wedge c_3 = \text{medium} \xrightarrow{g_c=0} d = \text{bad}$	1	$g_c = 0$
$c_1 = \text{low} \wedge c_3 = \text{high} \xrightarrow{g_c=3} d = \text{good}$	4	$g_c = 3$

Table 8.5 Classification Results Based on the Hybrid of VPRS and Probabilistic Neural Network

Actual Classification Results	*Classification Results of Probabilistic Neural Network Trained by Source Dataset*	*Classification Results of Probabilistic Neural Network Trained by the Dataset Reduced by β-Reduct*
Bad	Bad	Bad
Good	Good	Good
Good	Good	Good
Bad	Bad	Bad
Bad	Bad	Bad
Good	Good	Good
Good	Good	Good
Bad	Good	Good
Bad	Bad	Bad
Good	Good	Good

From the data in Table 8.5, we can see that the simulation classification qualities obtained by taking, respectively, the source datasets and the datasets with β-reduct as training sample sets to train probability neural network are identical, and the classification accuracies for both are 90 percent. Obviously, the VPRS method can identify the relatively important input data in probabilistic neural network and eliminate the redundant attributes in information system without losing almost any information; thus, it reduces the volume of the input data for the probabilistic neural network.

8.3 Transport Scheme Choice Decision Undertaking Consideration into Preference Information

8.3.1 Choice Decision Based on the Dominance Rough Set

In Table 8.1, c_1, c_2, c_3, c_4, and d can be regarded as preferential attributes. For c_1, the lower the transport cost is, the better, low > high; for c_2, the higher the batch scale is, the better, high > low; for c_3, the lower the inventory volume cost is, the better, low > high; for c_4, the quicker the response time is, the better, quick > slow; for d, good > bad.

Based on the dominance relation, the universe is portioned according to decision class into the following equivalence classes:

$$\frac{U}{C} = \{X_1, X_2, X_3, X_4, X_5\}$$

where $X_1 = \{n_1\}$, $X_2 = \{n_2\}$, $X_3 = \{n_3\}$, $X_4 = \{n_4\}$, $X_5 = \{n_5\}$, $X_6 = \{n_6, n_7, n_8, n_{10}\}$, $X_7 = \{n_9\}$.

$$\frac{U}{D} = \{Cl_1, Cl_2\}$$

where $Cl_1 = \{n_2, n_3, n_6, n_7, n_{10}\}$, $Cl_2 = \{n_1, n_4, n_5, n_9\}$.

Suppose that $Cl_1^{\leq}$ is the scheme of at most "bad" and $Cl_2^{\geq}$ is the scheme of at least "good." Because only two classes are considered, there are $Cl_1^{\leq} = Cl_1(\text{bad})$ and $Cl_2^{\geq} = Cl_2^{\geq}(\text{good})$.

Then, the lower approximation $\underline{apr}_P(Cl_1^{\leq})$ of $Cl_1^{\leq}$ and the upper approximation $\overline{apr}_P(Cl_1^{\leq})$ of $Cl_1^{\leq}$, the lower approximation $\underline{apr}_P(Cl_2^{\geq})$ of $Cl_2^{\geq}$ and the upper approximation $\overline{apr}_P(Cl_2^{\geq})$ of $Cl_2^{\geq}$ are, respectively,

$$\underline{apr}_P\left(Cl_1^{\leq}\right) = \{\{n_1\}, \{n_4\}, \{n_9\}\}$$

Table 8.6 Minimal Preferential Probability Rule Set for Reduct $\{c_1, c_3\}$

Rules	*Support*	*Confidence (Percent)*
$c_3 \geq \text{low} \xrightarrow{\text{100 percent}} d \in Cl_2$	2	100
$c_1 \leq \text{high} \wedge c_3 \leq \text{medium} \xrightarrow{\text{100 percent}} d \in Cl_1$	3	100

$$\overline{apr}_P\left(Cl_1^{\leq}\right) = \{\{n_1\},\{n_4\},\{n_9\},\{n_5,n_6,n_7,n_8,n_{10}\}\}$$

$$\underline{apr}_P\left(Cl_2^{\geq}\right) = \{\{n_2\},\{n_3\},\{n_5,n_6,n_7,n_8,n_{10}\}\}$$

$$\overline{apr}_P\left(Cl_2^{\geq}\right) = \{\{n_2\},\{n_3\},\{n_5,n_6,n_7,n_8,n_{10}\}\}$$

Based on the dominance-based rough set, the minimum reduct $\{c_1,c_3\}$ is computed, and the minimal preferential decision rules induced from the reduct $\{c_1,c_3\}$ are shown in Table 8.6.

According to the dominance relation, contractor n_5 is better than contractor n_{10}. Because as to the three preferential condition attributes, contractor n_5 is at least as good as contractor n_{10}, but contractor n_5 is worse than contractor n_{10} in the comprehensive evaluation, and the inconsistency can be approximately interpreted as incompatibility based on the dominance relation, although it cannot be approximately identified based on the indiscernibility relation. In the sense of indiscernibility relation, contractor n_5 and contractor n_{10} can be discerned.

8.3.2 Choice Decision Based on the Dominance-Based VPRS

According to decision class, the universe of Table 8.1 is portioned into the following equivalence classes by means of dominance relation:

$$\frac{U}{C} = \{X_1, X_2, X_3, X_4, X_5\}$$

where $X_1 = \{n_1\}$, $X_2 = \{n_2\}$, $X_3 = \{n_3\}$, $X_4 = \{n_4\}$, $X_5 = \{n_5\}$, $X_6 = \{n_6,n_7,n_8,n_{10}\}$, $X_7 = \{n_9\}$.

$$\frac{U}{D} = \{Cl_1, Cl_2\}$$

where $Cl_1 = \{n_2,n_3,n_6,n_7,n_{10}\}$, $Cl_2 = \{n_1,n_4,n_5,n_9\}$.

Let β = 60 percent, β-lower approximation $\underline{apr}_P^{\beta}(Cl_1^{\leq})$ of $Cl_1^{\leq}$ and β-upper approximation $\overline{apr}_P^{\beta}(Cl_1^{\leq})$ of $Cl_1^{\leq}$, β-lower approximation $\underline{apr}_P^{\beta}(Cl_2^{\geq})$ of $Cl_2^{\geq}$ and β-upper approximation $\overline{apr}_P^{\beta}(Cl_2^{\geq})$ of $Cl_2^{\geq}$, are, respectively, as follows:

$$\underline{apr}_P^{\beta}\left(Cl_1^{\leq}\right)=\{\{n_1\},\{n_4\},\{n_9\}\}\ \underline{apr}_P^{\beta}\left(Cl_1^{\leq}\right)=\{\{n_1\},\{n_4\},\{n_9\}\}$$

$$\overline{apr}_P^{\beta}\left(Cl_1^{\leq}\right)=\{\{n_1\},\{n_4\},\{n_9\}\}\ \overline{apr}_P^{\beta}\left(Cl_1^{\leq}\right)=\{\{n_1\},\{n_4\},\{n_9\}\}$$

$$\underline{apr}_P\left(Cl_2^{\geq}\right)=\{\{n_2\},\{n_3\},\{n_5,n_6,n_7,n_8,n_{10}\}\}$$

$$\overline{apr}_P\left(Cl_2^{\geq}\right)=\{\{n_2\},\{n_3\},\{n_5,n_6,n_7,n_8,n_{10}\}\}$$

The minimal β-reduct $\{c_1,c_3\}$ based on the dominance-based VPRS is computed, and the minimal preferential probabilistic rule set induced from the β-reduct $\{c_1,c_3\}$ is shown in Table 8.7.

$c_1 = \text{low} \wedge c_3 = \text{middle} \xrightarrow{100\,\text{percent}} d = \text{bad}$, $c_1 = \text{low} \wedge c_3 = \text{high} \xrightarrow{75\,\text{percent}} d = \text{good}$ in Table 8.7, that is to say, if transport cost is "low" and inventory cost is "medium", then the comprehensive evaluation on transport schemes is "bad" with the confident degree 100 percent. Nevertheless, if transport cost is "low" and inventory cost is "high," then the comprehensive evaluation on transport schemes is "good" with the confident degree 75 percent. Obviously, such rules are not reasonable, because the VPRS based on the indiscernible relation cannot capture the inconsistency coming from preference attributes. Such information with the extended VPRS approach based on the dominance relation could be considered as the inconsistency such as the rule $c_1 = \text{low} \wedge c_3 = \text{middle} \xrightarrow{60\,\text{percent}} d \in Cl_2$ in Table 8.7 avoids the unreasonable rules being generated by the inconsistencies

Table 8.7 Minimal Preferential Probabilistic Rule Set for β-Reduct $\{c_1, c_3\}$

Rules	*Support*	*Confidence (Percent)*
$c_3 \geq \text{low} \xrightarrow{100\,\text{percent}} d \in Cl_2$	2	100
$c_1 \leq \text{high} \wedge c_3 \leq \text{medium} \xrightarrow{100\,\text{percent}} d \in Cl_1$	3	100
$c_1 \geq \text{low} \wedge c_3 \leq \text{medium} \xrightarrow{60\,\text{percent}} d \in Cl_2$	5	60

from preference information. Furthermore, the preferential probability decision rules can be generated from the decision table with the preference information. Compared with Table 8.3, the rules in Table 8.7 have smaller number of rules than the ones in Table 8.3.

From Table 8.3, we can see that managers must consider inventory cost when choosing transport schemes. If the inventory can be cut down significantly, the transport scheme with high transport cost may be reasonable. Though the transport cost is low, the high inventory cost may result in very high total cost; such transport schemes may possibly be unreasonable. On the condition of noise and inconsistency preference information, we can obtain the preferential probability rules with the VPRS approach based on the dominance relation that are more reasonable than those with the VPRS approach based on the indiscernible relation.

Bibliography

Agrawal R, Imielinski T, Swami A. Mining association rules between sets of item in large databases. In: *Proceedings of ACM SIGMOD Conference on Management of Data*, Washington, DC, 1993, pp. 207–216.

Alicja M R, Leszek R. Fuzziness in information systems. *Electronic Notes in Theoretical Computer Science*, 2003, 82(4): 146–173.

Altman E. *Corporate Financial Distress*. New York: Wiley, 1983, pp. 62–75.

An A, Shan N, Chan C et al. Discovering rules for water demand prediction: An enhanced rough-set approach. *Engineering Application and Artificial Intelligence*, 1996, 9(6): 645–653.

Ananthanarayana V S, Murty M N, Subramanian D K. Tree structure for efficient data mining using rough sets. *Pattern Recognition Letters*, 2003, 24: 851–862.

Andrews R, Diederich J, Tickle A B. Survey and critique of techniques for extracting rules from trained artificial neural networks. *Knowledge-Based Systems*, 1995, 8(6): 373–389.

Asharaf S, Murty M N. A rough fuzzy approach to web usage categorization. *Fuzzy Sets and Systems*, 2004, 148: 119–129.

Banerjee M, Pal S K. Roughness of a fuzzy set. *Information Sciences*, 1996, 93: 235–246.

Banerjee M, Mitra S, Pal S K. Rough fuzzy MLP: Knowledge encoding and classification. *IEEE Transactions of Neural Networks*, 1998, 9(6): 1203–1216.

Bazan J G. A comparison of dynamic and non-dynamic rough set methods for extracting laws from decision tables. In: Polkowski L, Skowron A, Eds. *Rough Sets in Knowledge Discovery 2: Applications, Case Studies and Software Systems*. Heidelberg, Germany: Physica-Verlag, 1998, pp. 396–421.

Bazan J G, Skowron A, Synak P. Dynamic reducts as a tool for extracting laws from decision tables. In: Ras Z W, Zemenkova M, Eds. *Proceedings of the Eighth International Symposium*. New York, 1994, pp. 346–355.

Besdek J C, Pal S K. *Fuzzy Models for Pattern Recognition*. New York: IEEE Press, 1992.

Beynon M J, Dreffield N. An illustration of variable precision rough sets model: An analysis of the findings of the UK monopolies and mergers commission. *Computers and Operations Research*, 2005, 32: 1739–1759.

Beynon M, Peel M. Variable precision rough set theory and data discretization: An application to corporate failure prediction. *Omega*, 2001, 29: 561–576.

Bezdek J C. *Pattern Recognition with Fuzzy Objective Functions*. New York: Plenum, 1981.

Bezdek J C. The thirsty traveler visits gamont: A rejoinder to "comments on fuzzy set-what are they and why?". *IEEE Transactions on Fuzzy Systems*, 1994, 2(1): 43–45.

Bhatt R B, Gopal M. On fuzzy-rough sets approach to feature selection. *Pattern Recognition Letters*, 2005, 26(7): 965–975.

Bhatt R B, Gopal M. On the extension of functional dependency degree from crisp to fuzzy partitions. *Pattern Recognition Letters*, 2006, 27: 487–491.

Bose N K, Liang P. *Neural Network Fundamentals with Graphs, Algorithms and Applications*. New York: McGraw-Hill Inc., 1996, pp. 23–96.

Chan C C. A rough set approach to attribute generalization in data mining. *Journal of Information Sciences*, 1998, 107: 169–176.

Chen M S, Han J, Yu P S. Data mining: A overview from a database perception. *IEEE Transactions on Knowledge and Data Engineering*, 1996, 8(6): 866–883.

Chmielewski M R, Grzymala-Busse J W. Global discretization of continuous attributes as preprocessing for machine learning. *International Journal of Approximate Reasoning*, 1996, 15: 319–331.

Chopra S, Meindl P. *Supply Chain Management*, 2nd edn. Li L et al. (Trans.) Beijing, China: Social Sciences Documentation Press, 2003, pp. 287–331.

Chuang E, Lirong J. *Multimedia Computer Technology*, Shanghai, China: Shanghai Jiaotong University Press, 2003.

Cios K J, Pedrycz W, Swiniarski R. *Data Mining Methods for Knowledge Discovery*. Dordrecht, the Netherlands: Kluwer, 1998.

Collins E, Ghosh S, Scofield C. An application of a multiple neural network learning system to emulation of mortgage underwriting judgments. In: *Proceedings of the IEEE International Conference on Neural Networks*, San Diego, CA, 1988, pp. 459–466.

Creco S, Inuiguchi M, Slowinski R. Fuzzy rough sets and multiple-premise gradual decision rules. *International Journal of Approximate Reasoning*, 2006, 41: 179–211.

Criffiths B, Beynon M J. Expositing stages of VPRS analysis in an expert system: Application with bank credit ratings. *Expert Systems with Applications*, 2005, 29(4): 879–888.

Curram S, Mingers P J. Neural networks, decision tree induction and discriminant analysis: An empirical comparison. *Journal of the Operational Research Society*, 1994, 45(4): 440–450.

Dai J, Li Y. A heuristic genetic algorithm of attribute reduction in rough set. *Journal of Xi'an Jiaotong University*, 2002, 36(12): 1286–129.

Deng J. Grey system theory and a number of issues for application Progress [A]. Liu S, Xu Z X, Eds. *New Development of Grey System*. Wuhan, China: Huazhong University of Science and Technology Press, 1996.

Deng J. *Grey Prediction and Decision*. Wuhan, China: Huazhong University of Science and Technology Press, 2002.

Deng J. *Basic Method of Grey System*, 2nd edn. Wuhan, China: Huazhong University of Science and Technology Press, 2005.

Dhar V, Tuzhilin A. Abstract-driven pattern discovery in databases. *IEEE Transactions on Knowledge and Data Engineering*, 1993, 5(6): 926–937.

Diday E. *New Approaches in Classification and Data Analysis*. Berlin, Germany: Springer, 1994, pp. 482–488.

Dougherty J, Kohavi R, Sahami M. Supervised and unsupervised discretization of continuous features. In: Prieditis A, Russell S, Eds. *Proceedings of the 12th International Conference*. San Francisco, CA: Morgan Kaufmann, 1995, pp. 194–202.

Drakopoulos J A, Abdulkader A. Training neural networks with heterogeneous data. *Neural Networks*, 2005, 18: 596–601.

Dubois D, Prade H. *Fuzzy Sets and Systems*. New York: Academic Press, 1980, pp. 23–68.

Dubois D, Prade H. Rough fuzzy sets and fuzzy rough sets. *International Journal of General Systems*, 1990, 17(2–3): 191–209.

Dubois D, Prade H. Putting rough sets and fuzzy sets together. In: Slowinski R, Ed. *Intelligent Decision Support: Handbook of Applications and Advances of the Rough Sets Theory*. Boston, MA: Kluwer, 1992, pp. 204–232.

Dubois D, Prade H. *The Oriental Aspect of Reasoning about Data*. London, U.K.: Academic Publishers, 1999.

Fan T F, Liu D R, Tzeng G H. Rough set-based logics for multicriteria decision analysis. *European Journal of Operational Research*, 2006, 121: 1–16.

Fang M, Zhao X, Sun H. A rough fuzzy neural network classifier and its application. *Journal of Hefei University of Technology*, 2005, 28(9): 1058–1061.

Fayyad U M, Irani K B. Multi-interval discretization of continuous-valued attributes for classification learning. In: *Proceedings of the 13th International Joint Conference on Artificial Intelligence*. Los Altos, CA: Morgan Kaufmann, 1993, pp. 1022–1027.

Fayyad U M, Piatetsky-Shapiro, Smyth P et al. *Advances in Knowledge Discovery and Data Mining*. Menlo Park, CA: AAAI/MIT Press, 1996.

Feng J, Tang R, Gao L. A survey on grey reasoning and its intelligent application. *Computer Engineering and Science*, 2006, 28(3): 131–133.

Fletcher D, Goss E. Forecasting with neural networks: An application using bankruptcy data. *Information and Management*, 1993, 24: 159–167.

Fodor J, Roubens M. *Fuzzy Preference Modeling and Multicriteria Decision Support*. Dordrecht, the Netherlands: Kluwer, 1994, pp. 120–230.

Gilbert L R, Menon K, Schwartz K B. Predicting bankruptcy for firms in financial distress. *Journal of Business Finance and Accounting*, 1990, 17(1): 161–171.

Glorfeld L. Methodology of simplification and interpretation of back propagation based neural network models. *Expert Systems with Applications*, 1996, 10(1): 37–54.

Goh C, Law R. Incorporating the rough sets theory into travel demand analysis. *Tourism Management*, 2003, 24: 511–517.

Golden R M. *Mathematical Methods for Neural Network Analysis and Design*. Cambridge, MA: MIT Press, 1996, pp. 86–112.

Greco S, Matarazzo B, Slowinski R. Multiple criteria decision making. In: *Proceedings of the 12th International Conference*. Berlin, Germany: Springer, 1997a, pp. 318–329.

Greco S, Matarazzo B, Slowinski R. Rough set approach to multi-attribute choice and ranking problems. In: Fandel G, Gal T, Eds. *Proceedings of the Twelfth International Conference*. Berlin, Germany: Springer, 1997b, pp. 318–329.

Greco S, Matarazzo B, Slowinski R. Rough approximation of a preference relation by dominance relations. *European Journal of Operational Research*, 1999, 117(1): 63–83.

Greco S, Matarazzo B, Slowinski R. Rough set processing of vague information using fuzzy similarity relations. In: Calude C S, Paun G, Eds. *Finite versus Infinite Contributions to an Eternal Dilemma*. London, U.K.: Springer, 2000, pp. 149–173.

Greco S, Matarazzo B, Slowinski R. Rough sets theory for multicriteria decision analysis. *European Journal of Operational Research*, 2001, 129: 1–47.

Greco S, Matarazzo B, Slowinski R, Tsoukias A. Exploitation of a rough approximation of the outranking relation in multicriteria choice and ranking. In: Stewart T J, Honert R C, Eds. *Trends in Multicriteria Decision Making*. Berlin, Germany: Springer, 1998, pp. 450–460.

Grzymala-Busse J W. LERS—A system for learning from examples based on rough sets. In: Slowinski R, Ed. *Intelligent Decision Support*. Dordrecht, the Netherlands: Kluwer, 1992, pp. 3–18.

Grzymala-Busse J W. A new version of the rule induction system LERS. *Fundamenta Informaticae*, 1997, 31: 27–39.

Grzymala-Busse J W. LERS—A knowledge discovery system. In: Polkowski L, Skowron A, Eds. *Rough Sets in Knowledge Discovery*. Wurzburg, Germany: Physics, 1998, pp. 562–565.

Gunn J D, Grzymala-Busse J W. Global temperature stability by rule induction: An interdisciplinary bridge. *Human Ecology*, 1994, 22(1): 59–81.

Hashemi R, Blanc L L A, Rucks C T et al. A hybrid intelligent system for predicting bank holding structures. *European Journal of Operational Research*, 1998, 109(2): 390–402.

Hassoun M F. *Fundamentals of Artificial Neural Networks*. Cambridge, MA: MIT Press, 1995, pp. 33–67.

Hayki S. *Neural Networks—A Comprehensive Foundation*. New York: Macmillan College Publishing Company, 1994, pp. 26–79.

He Z. *Fuzzy Mathematics and Its Applications*. Tianjin, China: Tianjin Science and Technology Publishing House, 1985, pp. 121–125.

He Y, Hu S. Preferred sites of the air force air material supply point based on rough set theory. *Systems Engineering Theory and Practice*, 2003, 7: 95–99.

Herta J, Krogh A, Palmer R G. *Introduction to the Theory of Neural Computation*. Cambridge, MA: Addison-Wesley, 1991, pp. 11–47.

Höhle U. Quotients with respect to similarity relations. *Fuzzy Sets and Systems*, 1988, 27: 31–44.

Hong T P, Wang T T, Wang S L. Learning a coverage set of maximally general fuzzy rules by rough sets. *Expert Systems with Applications*, 2000, 19: 100–102.

Hu X, Cercone N. Learning in relational database: A rough set approach. *Computational Intelligence*, 1995, 11(2): 323–338.

Huang C C, Tseng T L. Rough set approach to case-based reasoning application. *Expert Systems with Applications*, 2004, 26: 369–385.

Huynh V N, Nakamori Y. A roughness measure for fuzzy sets. *Information Sciences*, 2005, 173: 255–275.

Inuiguchi M, Tanino T. Fuzzy rough sets based on certainty qualifications. In: *Proceedings of the Fourth AFSS Symposium*. Tsukuba, Japan, 2000, pp. 433–438.

Janusz A S, Dale E N, Kirk S. A mathematical foundation for improved reduct generation in information systems. *Knowledge and Information Systems*, 2000, 2(2): 131–146.

Jelonek J. Rough set reduction of attributes and their domains for neural networks. *Computational Intelligence*, 1995, 11(2): 323–338.

Jiang Y, Lou Z, Li M. Cutting parameter optimization of NC machining based on fuzzy rough set theory. *Journal of Shanghai Jiaotong University*, 2005, 39(7): 1115–1118.

Jing L, Da Q, Chen W. Variable precision rough set and a fuzzy measure of knowledge based on variable precision rough set. *Journal of Southeast University (English)*, 2002, 18(4): 351–355.

Kandel A. *Fuzzy Techniques in Pattern Recognition*. New York: Wiley, 1982.

Kandel A. *Fuzzy Mathematical Techniques with Applications*. Reading, MA: Addison-Wesley, 1986, pp. 110–200.

Katzberg J D, Ziarko W. Variable precision extension of rough sets. *Fundamental Information*, 1996, 27: 155–168.

Kerber R. ChiMerge: Discretization of numeric attributes. In: *Proceedings of the Ninth International Conference on Artificial Intelligence*. Cambridge, MA: The MITS Press, 1992, pp. 123–128.

Kidd A. *Knowledge Acquisition for Expert Systems: A Practical Handbook*. New York: Plenum Press, 1987, pp. 123–167.

Klir G J, Folger T A. *Fuzzy Sets, Uncertainty and Information*. Englewood Cliffs, NJ: Prentice-Hall, 1988.

Komorowski J, Pawlak Z, Polkowski L et al. Rough sets: A tutorial. In: Pal S K, Skowron A, Eds. *Rough Fuzzy Hybridization: A New Trend in Decision Making*. Singapore: Springer, 1999, pp. 1–98.

Kryszkiewicz M, Rybinski H. Finding reducts in composed information systems(C). In: Ziarko W, Ed. *Rough Sets, Fuzzy Sets and Knowledge Discovery*. Holland: Elsevier, 1993, pp. 261–273.

Kryszkiewicz M, Rybinski H. Computation of reducts of composed information systems. *Fundamental Information*, 1996, 27: 183–195.

Kudo M, Sklansky J. Comparison of algorithms that select features for pattern classifiers. *Pattern Recognition*, 2000, 33(1): 25–41.

Li J, Fan X, Huang P. Knowledge of theory and its application based on rough sets theory. *Systems Engineering Theory Methodology Applications*, 2001, 10(3): 184–188.

Li K, Liu Y S. Rough set based attribute reduction approach in data mining. In: *Proceedings of 2002 International Conference on Machine Learning and Cybernetics*. Piscataway, NJ: IEEE Press, 2002, pp. 60–63.

Li R, Wang Z O. Mining classification rules using rough sets and neural networks. *European Journal of Operational Research*, 2004, 157: 439–448.

Lin C T, Lee C S G. *Neural Fuzzy Systems*. Englewood Cliffs, NJ: Prentice Hall, 1996, pp. 23–78.

Lingras P J, Yao Y Y. Data mining using extensions of the rough set model. *Journal of the American Society for Information Science*, 1998, 49(5): 415–422.

Lirong J. Failure prediction of listed companies based on rough sets. *Computer Application Research (Supplement)*, 2003, 1: 91–93.

Lirong J. *Introduction of e-Commerce*. Beijing, China: Science Press, 2005.

Lirong J. Factor analysis for the impact of e-commerce market participants earnings. *Business Studies*, 2006, 3(9): 164–167.

Lirong J. Enterprise E-supply chain integration challenges and strategies based on Internet. *Business Studies*, 2007, 10: 207–211.

Lirong J, Bai Q. A survey on the theory and methodology of rough set. *Journal of Zhengzhou Institute of Aeronautical Industry Management*, 2005, 23(3): 27–32.

Lirong J, Liu S. Extension of rough set methodology for probabilistic decision analysis from preferential multiple attribute decision tables. *Journal of Nanjing University of Aeronautics and Astronautics*, 2005a, 37(4): 270–275.

Lirong J, Liu S. Extension of variable precision rough sets methodology for probabilistic rules from fuzzy decision table. *Control and Decision*, 2005b, 20(11): 1291–1295.

Lirong J, Liu S. Knowledge discovery methods based on the hybrid of VPRS and PNN. *Intelligence Journal*, 2005c, 24(4): 426–432.

Lirong J, Liu S. A survey on building the system of project management capacity. *Industrial Techno-Economic*, 2006a, 25(9): 108–112.

Lirong J, Liu S. Extension of rough sets model for probabilistic decision analysis from rough-fuzzy decision tables. *Journal of Southeast University*, 2006b, 2(5): 246–250.

Lirong J, Liu S. The Definition of Grey Degree of Grey Number Based on Rough Membership Function and Grey Rough Approximation. Grey system theory and application, Chinese higher science and technology center, 2006c, pp. 232–240.

Lirong J, Da Q, Chen W. Variable precision rough set and a fuzzy measure of knowledge based on variable precision rough set. *Journal of Southeast University*, 2002, 18(4): 351–355.

Lirong J, Da Q, Chen W. A method of rule induction based on rough set theory in inconsistent information system. *Management Science in China*, 2003, 11(4): 91–96.

Lirong J, Da Q, Chen W. A hierarchical granulation of knowledge based on variable precision rough set theory. *Journal of Management Engineering*, 2004, 18(2): 60–64.

Lirong J, Li M, Liu L. Post-doctoral theses collection of Jiangsu Province (School of Humanities and Social and Economic Management Volume), Personnel Office of Jiangsu Province, 2005, pp. 120–124.

Lirong J, Liu S, Hu C et al. Exploration and analysis for comprehensive budget management practice of Lianxu company. *Modern Management Science*, 2006, 12(1): 15–17.

Lirong J, Liu S, Fang Z et al. University teaching and research building performance evaluation based on dominance rough sets theory. *Journal of Management Engineering*, 2007, 21(3): 132–137.

Lisboa. Xing C et al. *Application of Modern Neural Network*. Beijing, China: Electronics Industry Press, 1996, pp. 1–30.

Liu S. On measure of grey information. *The Journal of Grey System*, 1995, 7(2): 97–101.

Liu S. Axioms on grey degree. *The Journal of Grey System*, 1996, 8(4): 396–400.

Liu Z. *Modern Project Management*. Tianjin, China: Tianjin University Press, 1997, p. 209.

Liu S. Emergence and development of the grey system theory. *Journal of Nanjing University of Aeronautics and Astronautics*, 2004, 36(2): 267–272.

Liu S, Lin Y. *An Introduction to Grey Systems: Foundations, Methodology and Applications*. Slippery Rock, PA: IIGSS Academic Publisher, 1998.

Liu S, Lin Y. *Grey Information: Theory and Practical Applications*. London, U.K.: Springer-Verlag London Ltd., 2005.

Liu S, Zhao L, Wang Z. A new method of evaluation for venture capital. *Management Science in China*, 2001, 9(2): 22–26.

Liu S, Dang Y, Fang Z. *Grey System Theory and Its Application*, 3rd edn. Beijing, China: Science Press, 2004.

Liu S, Fang Z, Gong Z. A survey on a new characterization of interval grey number and its algorithms. *The Ninth Annual Meeting*, Systems Engineering Institute of Jiangsu Province, Jiangsu, China, 2, pp. 519–523.

Liu S, Li B, Dang Y. The G-C-D model and technical advance. *Kybernetes: The International Journal of Systems and Cybernetics*, 2004, 33(2): 303–309.

Liu S, Li N, Dang Y. *Econometrics*. Nanjing, China: Southeast University Press, 2006.

Liu S, Wu H, Lirong J. *Applied Statistics*. Beijing, China: Higher Education Press, 2007.

Lu X, Chen S, Wu J. A rough set knowledge reduction algorithm based on genetic algorithm. *Computer Engineering*, 2003, 29(1): 56–59.

Luo Y. Discussion on university discipline construction. *Research on Higher Education*, 2005, 26(7): 45–50.

Luo D, Liu S, Wu S. A research on the combined decision model based on the grey theory and rough sets. *Journal of Xiamen University*, 2004, 43(1): 26–30.

McSherry D. Knowledge discovery by inspection. *Decision Support Systems*, 1997, 21: 43–47.

Miao D, Hu G. Knowledge reduction of a heuristic algorithm. *Computer Research and Development*, 1999, 36(6): 681–684.

Michalski R S, Carbonell J G, Mitcell T M. *Machine Learning: An Artificial Intelligence Approach*. Palo Alto, CA: Tioga, 1983, pp. 83–134.

Michalski R S, Bratko I, Kubat M. *Machine Learning and Data Mining—Methods and Applications*. New York: Wiley, 1998, pp. 62–96.

Mienko R, Slowinski R, Stefanowski J et al. Rough family-software implementation of rough set based data analysis and rule discovery techniques. In: Tsumoto S, Ed. *Proceedings of the Fourth International Workshop on Rough Set*. Tokyo, Japan: Tokyo University Press, 1996, pp. 437–440.

Morris R. *Early Warning Indicators of Corporate Failure*. Aldershot, U.K.: Ashgate, 1997, pp. 56–72.

Nguyen H S, Szczuka M, Slezak D. Neural network design: Rough set approach to real-valued data. In: *Proceedings of the First European Symposium Principles of Data Mining and Knowledge Discovery*. Trondheim, Norway, 1997, pp. 359–366.

Ning X, Liu S. *Management Forecast and Decision Method*. Beijing, China: Science Press, 2003.

Øhrn A, Komorowski J, Skowron A et al. The design and implementation of a knowledge discovery toolkit based on rough sets-the ROSETTA system. In: Polkowski L, Skowron A, Eds. *Rough Sets in Knowledge Discovery 1: Methodology and Applications*. Heidelberg, Germany: Physica, 1998, pp. 376–399.

Ohsuga S. Symbol processing by non-symbol processor. In: *Proceedings of Fourth Pacific Rim International Conference on Artificial Intelligence*. Cairns, Australia, 1996, pp. 193–205.

Ovchinnikov S V. Similarity relations fuzzy partitions and fuzzy orderings. *Fuzzy Sets and Systems*, 1991, 40: 107–126.

Pal S K. Soft data mining, computational theory of perceptions, and rough-fuzzy approach. *Information Sciences*, 2004, 163: 5–12.

Pal S K, Majumder D D. *Fuzzy Mathematical Approach to Pattern Recognition*. New York: Wiley, 1986, pp. 110–256.

Pal S K, Pedryzc W, Skowron A et al. Rough-neuro computing. *Neuro Computing*, 1996, 36: 1–4.

Pal S K, Shankar B U, Mitra P. Granular computing, rough entropy and object extraction. *Pattern Recognition Letters*, 2005, 26(16): 2509–2517.

Pan Y, Lu J, Lirong J et al. *Database of e-Government*. Beijing, China: Beijing University Press, 2005.

Pattaraintakorn P, Cercone N, Naruedomkul K. Rule learning: Ordinal prediction based on rough sets. *Applied Mathematics Letters*, 2006, 19(12): 1300–1307.

Pattaraintakorn P, Cercone N, Naruedomkul K. Rule learning: Ordinal prediction based on rough sets and soft-computing. *Applied Mathematics Letters*, 2006, 19: 1300–1307.

Pawlak Z. Rough classification. *International Journal of Man-Machine Studies*, 1984, 20: 469–483.

Pawlak Z. Vagueness and uncertainty: A rough set perspective[R]. Technical Report ICS Research Report, Warsaw, Poland, 1994, pp. 23–65.

Pawlak Z. Rough set approach to knowledge-based decision support. *European Journal of Operational Research*, 1997, 99: 48–57.

Pawlak Z. Granularity of knowledge, indiscernibility and rough sets. In: *Proceedings of 1998 IEEE International Conference on Fuzzy Systems*. Anchorage, AK, 1998a, pp. 106–110.

Pawlak Z. Rough sets and decision analysis. In: *Proceedings of the Fifth IIASA Workshop on Decision Analysis and Support*. Laxenburg, Austria, 1998b, pp. 123–127.

Pawlak Z. Rough sets. *International Journal of Information and Computer Sciences*, 1998c, 49(5): 415–422.

Pawlak Z. Rough sets and intelligent data analysis. *Information Science*, 2002, 147: 1–12.

Pawlak Z. Some remarks on conflict analysis. *European Journal of Operational Research*, 2005, 166: 649–654.

Pawlak Z, Skowron A. Rough sets and boolean reasoning. *Information Sciences*, 2007, 177: 41–47.

Pawlak Z, Slowinski R. Rough set approach to multi-attribute decision analysis. *European Journal of Operational Research*, 1994, 72: 443–459.

Peel M. *The Liquidation/Merger Alternative*. Aldershot, U.K.: Avebury, 1990, pp. 102–135.

Polkowski L, Skowron A. Rough methodology. In: Charlotte N C, Ed. *Proceedings of the Symposium on Methodologies for Intelligent Systems*. Berlin, Germany: Springer Verlag, 1994, pp. 85–94.

Polkowski L, Skowron A. *Rough Sets and Current Trends in Computing*. Berlin, Germany: Springer, 1998a, pp. 283–298.

Polkowski L, Skowron A. *Rough Sets in Knowledge Discovery*. Heidelberg, Germany: Physica, 1998b, pp. 422–450.

Pomerol J C. Artificial intelligence and human decision making. *European Journal of Operational Research*, 1997, 99: 3–25.

Predki B, Slowinski R, Stefanowski J et al. ROSE-software implementation of the rough set theory. In: Polkowski L, Skowron A, Eds. *Rough Sets and Current Trends in Computing*. Berlin, Germany: Springer, 1998, pp. 605–608.

Questier F, Rollier I A, Walczak B. Application of rough set theory to feature selection for unsupervised clustering. *Chemometrics and Intelligent Laboratory Systems*, 2002, 63: 155–167.

Radzikowska A, Kerre E E. Characterization of main classes of fuzzy relations using fuzzy modal operators. *Fuzzy Sets and Systems*, 2005, 152: 223–247.

Rao D V, Sarma V V S. A rough-fuzzy approach for retrieval of candidate components for software reuse. *Pattern Recognition Letters*, 2003, 24: 875–886.

Richard O, Duda P E, Hart D G S. *Pattern Classification*. Li H D et al. (Trans.) Beijing, China: Machinery Industry Press, 2003, pp. 490–498.

Ripley B D. *Pattern Recognition and Neural Networks*. Cambridge, U.K.: Cambridge University Press, 1996, pp. 34–67.

Rong X. *Principles and Methods of Fuzzy Mathematics*. Jiangsu, China: Chinese Mining Industry University Press, 1999, pp. 44–47.

Roy B. The outranking approach and the foundation of ELECTRE methods. *Theory and Decision*, 1991, 31: 49–73.

Roy B. Decision science or decision aid science. *European Journal of Operational Research*, 1993, 66: 184–203.

Rumelhart D E, Hinton G E, Williams R J. Learning internal representation by error propagation. In: Rumelhart D E, Clelland M C, Eds. *Parallel and Distributed Processing*. Cambridge, MA: MIT Press, 1986, pp. 67–98.

Sarkar M. Rough-fuzzy functions in classification, 2002, 132: 353–369.

Seungkoo L, George V. Application of rough set theory to detection of automotive glass. *Mathematics and Computers in Simulation*, 2002, 60: 225–231.

Shen Q, Chouchoulas A. A rough-fuzzy approach for generating classification rules. *Pattern Recognition*, 2002, 35(11): 2425–2438.

Singh S, Dey L. A new customized document categorization scheme using rough membership. *Applied Soft Computing*, 2005, 5: 373–390.

Skowron A. Boolean reasoning for decision rules generation. In: Komorowski J, Ras Z W, Eds. *Methodologies for Intelligent Systems*. Berlin, Germany: Springer, 1993, pp. 293–308.

Skowron A, Polkowski L. Decision algorithms: A survey of rough set-theoretic methods. *Fundamental Information*, 1997, 27(3): 345–358.

Skowron A, Polkowski L. *Rough Sets in Data Mining and Knowledge Discovery*. Heidelberg, Germany: Physica, 1998, pp. 500–529.

Skowron A, Rauszer C. *Intelligent Decision Support—Handbook of Applications and Advances of the Rough Sets Theory*. Dordrecht, the Netherlands: Kluwer Academic Publishers, 1992a, pp. 331–362.

Skowron A, Rauszer C. The discernibility matrices and functions in information systems. In: Slowinski R, Ed. *Intelligent Decision Support Handbook of Applications and Advances of the Rough Sets Theory*. Dordrecht, the Netherlands: Kluwer Academic Publishers, 1992b, pp. 331–336.

Slezak D, Ziarko W. The investigation of the Bayesian rough set model. *International Journal of Approximate Reasoning*, 2005, 40: 81–91.

Slovic P. Choice between equally-valued alternatives. *Journal of Experimental Psychology: Human Perception Performance*, 1975, 1: 280–287.

Slowinski R. *Handbook of Applications and Advances of the Rough Set Theory*. Dordrecht, the Netherlands: Kluwer Academic Publishers, 1992a, pp. 233–269.

Slowinski R. *Intelligent Decision Support: Handbook of Applications and Advances of the Rough Sets Theory*. Dordrecht, the Netherlands: Kluwer Academic Publishers, 1992b, pp. 331–362.

Slowinski R, Stefanowski J. Rough classification in incomplete information systems. *Mathematical Computer Modeling*, 1989, 12(10): 1347–1357.

Slowinski R, Vanderpooten D. Similarity relation as a basis for rough approximations. In: Wang P P, Ed. *Advances in Machine Intelligence and Soft-Computing*. Durham, NC: Duke University Press, 1995, pp. 17–33.

Slowinski R, Vanderpooten D. A generalized definition of rough approximations based on similarity. *IEEE Transactions on Data and Knowledge Engineering*, 2000, 12(2): 331–336.

Slowinski R, Zopounidis C. Application of the rough set approach to evaluation of bankruptcy risk. *Finance and Management*, 1995, 3(4): 27–41.

Specht D F. Probabilistic neural networks and the polynomial Adeline as complementary techniques for classification. *IEEE Transactions on Neural Networks*, 1990a, 1(1): 111–121.

Specht D F. Probabilistic neural networks. *Neural Networks*, 1990b, 3(1): 109–118.

Srinivasam P, Ruiz M E, Kraft D H et al. Vocabulary mining for information retrieval: Rough sets and fuzzy sets. *Information Processing and Management*, 2001, 37: 15–38.

Stefanowski J. On rough set based approaches to induction of decision rules. In: Skowron A, Polkowski L, Eds. *Rough Sets in Data Mining and Knowledge Discovery*. Heidelberg, Germany: Physica, 1998, pp. 500–529.

Stepaniuk J, Kierzkowska K. Hybrid classifier based on rough sets and neural networks. *Electronic Notes in Theoretical Computer Science*, 2003, 82(4): 228–238.

Subramanian V, Hung M, Hu M. An experimental evaluation of neural networks for classification. *Computers and Operations Research*, 1993, 20(7): 769–782.

Supriya K D, Krishna P R. Clustering web transactions using rough approximation. *Fuzzy Sets and Systems*, 2004, 148(1): 131–138.

Susmaga R, Michalowski W, Slowinski R. Identifying regularities in stock portfolio tilting. *International Institute for Applied Systems Analysis*, Laxenburg, Austria, 1997, pp. 12–20.

Tang X, Liu S. Post-evaluation method for Government's investment projects. *The Ninth Annual Meeting*, Systems Engineering Institute of Jiangsu Province, Jiangsu, China, 2005, 2, pp. 450–455.

Tao Z, Xu B, Wang D, Li R. Rough set knowledge reduction algorithm based on genetic algorithm. *Systems Engineering*, 2003, 21(4): 116–122.

Tsau Y L, Ping Y. Heuristically fast finding of the shortest reducts. In: *Proceedings of 2004 International Conference on Rough Sets and Current Trends in Computing*. Berlin, Germany: Springer, 2004, pp. 465–470.

Tsumoto S. Knowledge discovery in medical databases based on rough sets and attribute-oriented generalization. In: *Proceedings of the IEEE FUZZ99*. Anchorage, AK: IEEE Press, 1999, pp. 1296–1297.

Vidyasagar M. *A Theory of Learning and Generalization with Application to Neural Networks and Control Systems*. Dordrecht, the Netherlands: Kluwer, 1996, pp. 33–73.

Vinterbo S, Øhrn A. Minimal approximate hitting sets and rules template. *International Journal of Approximate Reasoning*, 2000, 25(2): 123–143.

Walczak B, Massart D L. Rough sets theory, *Chemometrics and Intelligent Laboratory Systems*, 1999, 47: 1–16.

Wang F H. On acquiring classification knowledge form noisy data based on rough set. *Expert Systems Application*, 2005, 29: 49–64.

Wang H, Ma Y. Simulation of mine ventilation system reliability evaluation based on rough set—Neural networks. *Systems Engineering Theory and Practice*, 2005, 7: 82–86.

Wang W, Zhou D. A rough set knowledge reduction algorithm based on genetic algorithm. *Journal of System Simulation*, 2001, 13(Suppl): 91–94.

Wang Q, Wang F, Zuo Q et al. *Mathematical Basis of the Grey Theory*. Wuhan, China: Huazhong University of Science and Technology Press, 1996.

Wang Y, Zhang C, Shao H. Heuristic knowledge reduction algorithm research and application. *Control and Decision*, 2001, 16(6): 886–889.

Wen X, Zhou L, Li X et al. *Neural Network Simulation and Application Based on MATLAB*. Beijing, China: Science Press, 2003, pp. 289–291.

Wong S K M, Ziarko W. Comparison of the probabilistic approximate classification and the fuzzy set model. *Fuzzy Sets and Systems*, 1987, 21: 357–362.

Wroblewski J. Genetic algorithm in decomposition and classification problems(C). In: Polkowski L, Slowron A, Eds. *Rough Sets in Knowledge Discovery 2: Applications, Case Studies and Software Systems*. Heidelberg, Germany: Physica-Verlag, 1998, pp. 472–492.

Wu W Z, Mi J S, Zhang W X. Generalized fuzzy rough sets. *Information Sciences*, 2003, 151(1): 263–282.

Xiao J, Wu J, Yang S. The attribute reduction method of knowledge based on heuristic and its application in the evaluation system. *Systems Engineering*, 2002, 20(1): 92–96.

Xiao Y, Wu B, Qiu G et al. The problem of the university disciplinary construction management and thinking. *Research and Development Management*, 2006, 18(4): 127–130.

Xie J, Ye S. Reflections and recommendations for evaluation index system of liberal arts disciplines. *China's Higher Education Research*, 2005, 3(4): 26–30.

Yao Y Y. Combination of rough and fuzzy sets based on α-level sets. In: Lin T Y, Cerone N, Eds. *Rough Sets and Data Mining: Analysis for Imprecise Data*. Boston, MA: Kluwer, 1997, pp. 301–302.

Yao Y. A comparative study of fuzzy sets and rough sets. *Information Sciences*, 1998, 109: 227–242.

Yegnanarayana B. Artificial neural networks for pattern recognition. *Sadhana*, 1994, 19(1): 147–169.

Yin Y, Cao C, Zhang B. Classification rule discovery based on rough sets theory. *Journal of Chongqing University*, 2000, 23(1): 63–65.

Yu P, Lirong J, Da Q. Variable precision rough sets methodology for probabilistic rules discovery from a multi-criteria decision table. *Management Science in China*, 2005, 13(1): 95–100.

Zadeh L A. Fuzzy sets. *Information and Control*, 1965, 8: 338–353.

Zadeh L A. Fuzzy graphs, rough sets and information granularity. In: *Proceedings of the Third International Workshop on Rough Sets and Soft Computing*. San Jose, CA, 1994, pp. 210–212.

Zadeh L A. Toward a theory of fuzzy information granulation and its centrality in human reasoning and fuzzy logic. *Fuzzy Sets and Systems*, 1997a, 90: 111–127.

Zadeh L A. Towards a theory of fuzzy information granulation and its centrality in human reasoning and fuzzy logic. *Fuzzy Sets and Systems*, 1997b, 19: 111–127.

Zaras K. Rough approximation of a preference relation by a multi-attribute dominance for deterministic, stochastic and fuzzy decision problems. *European Journal of Operational Research*, 2004, 196: 206.

Zhai L Y, Khoo L P, Fok S C. Feature extraction using rough set theory and genetic algorithms-an application for the simplification of product quality evaluation. *Computers and Industrial Engineering*, 2002, 43(4): 661–676.

Zhang W, Wu W, Liang J. *Rough Set Theory and Methods*. Beijing, China: Science Press, 2001, pp. 123–131.

Zhang J, Cong G, Zhou Z. A IT project evaluation decision model based on rough sets and grey cluster. *Management Review*, 2005, 17(10): 29–33.

Zhong N, Skowron A, Ohsuga S. New direction in rough sets data mining and granular-soft computing. In: *Proceedings of the Seventh International Workshop*. Yamaguchi, Japan, 1999, pp. 8–21.

Ziarko W. Analysis of uncertain information in the framework of variable precision rough sets. *Foundations of Computing and Decision Sciences*, 1993a, 18: 381–396.

Ziarko W. Variable precision rough set model. *Journal of Computer and System Sciences*, 1993b, 46(1): 39–59.

Ziarko W P. *Rough Set, Fuzzy Sets and Knowledge Discovery*. London, U.K.: Springer, 1994, pp. 366–376.

Ziarko W. *Rough Sets, Fuzzy Sets and Knowledge Discovery*. Singapore: Springer, 1999, pp. 1–98.

Ziarko W P, Shan N. An incremental learning algorithm for constructing decision rules. In: Ziarko W P, Ed. *Rough Sets, Fuzzy Sets and Knowledge Discovery*. London, U.K.: Springer, 1994, pp. 326–334.

Zighed D, Rabaseda S, Rakotomala R. FUSINTER: A method for discretization of continuous attributes. *Fuzziness and Knowledge-Based Systems*, 1998, 6(3): 266–307.

Index

E

F

W

T - #0119 - 101024 - C0 - 234/156/15 [17] - CB - 9781420087482 - Gloss Lamination